Animal Welfare

Series Editor
Professor Clive Phillips
Foundation Chair of Animal Welfare
Centre for Animal Welfare and Ethics
School of Veterinary Science
University of Queensland
Gatton 4343, QLD
Australia

For further volumes:
http://www.springer.com/series/5675

Phil Glatz • Christine Lunam • Irek Malecki
Editors

The Welfare of Farmed Ratites

Springer

Editors
Dr. Phil Glatz
South Australian Research and
Development Institute
Roseworthy Campus
University of Adelaide
Roseworthy, 5371, SA
Australia
phil.glatz@sa.gov.au

Dr. Christine Lunam
Flinders University
Sensory Nervous System Laboratory
Department of Anatomy and Histology
Box GPO Box 21, Adelaide, 5001, SA
Australia
chris.lunam@flinders.edu.au

Dr. Irek Malecki
The University of Western Australia
School of Animal Biology
Faculty of Natural and Agricultural
Sciences
35 Stirling Highway, Crawley, 6009, WA
Australia
irek.malecki@uwa.edu.au

ISSN 1572-7408
ISBN 978-3-642-19296-8 e-ISBN 978-3-642-19297-5
DOI 10.1007/978-3-642-19297-5
Springer Heidelberg Dordrecht London New York

Library of Congress Control Number: 2011929641

Cover design: SPI Publisher Services

Printed on acid-free paper

Springer is part of Springer Science+Business Media (www.springer.com)

Animal Welfare Series Preface

Animal welfare is attracting increasing interest worldwide, especially in developed countries where the knowledge and resources are available to (at least potentially) provide better management systems for farm animals, as well as companion, zoo and laboratory animals. The key requirements for adequate food, water, a suitable environment, companionship and health are important for animals kept for all of these purposes.

There has been increased attention given to farm animal welfare in the West in recent years. This derives largely from the fact that the relentless pursuit of financial reward and efficiency, to satisfy market demands, has led to the development of intensive animal production systems that challenge the conscience of many consumers in those countries.

In developing countries, human survival is still a daily uncertainty, so that provision for animal welfare has to be balanced against human welfare. Animal welfare is usually a priority only if it supports the output of the animal, be it food, work, clothing, sport or companionship. In principle, the welfare needs of both humans and animals can be provided for, in both developing and developed countries, if resources are properly husbanded. In reality, however, the inequitable division of the world's riches creates physical and psychological poverty for humans and animals alike in many parts of the world. Livestock is the world's biggest land users (Food and Agriculture Organisation 2002) and the farmed animal population is increasing rapidly to meet the needs of an expanding human population. This results in a tendency to allocate fewer resources to each animal and to value individual animals less, for example, in the case of farmed poultry where flocks of over 30,000 meat birds and 50,000 laying hens are common. The largest layer farms have more than one million hens in cages 12 tiers high. In these circumstances, the importance of each individual's welfare is diminished.

Increased attention to welfare issues is just as evident for companion, laboratory, wild and zoo animals. Of increasing importance is the ethical management of breeding programmes, since genetic manipulation is more feasible, but there is less public tolerance of the deliberate breeding of animals for improved productivity if it comes at the expense of animal welfare. However, the quest for producing novel genotypes has fascinated breeders for centuries. Dog and cat breeders have produced a variety of extreme forms with adverse effects on their welfare, but nowadays the quest by breeders is most avidly pursued in the laboratory, where the mouse is genetically manipulated with equally profound effects.

The intimate connection between animals and humans that was once so essential for good animal welfare is rare nowadays, have been superseded by technologically efficient production systems where animals in farms and laboratories are tended by increasingly few humans in the drive to enhance labour efficiency. With today's busy lifestyle, companion animals too may suffer from reduced contact with humans, although their value in providing companionship, particularly for certain groups such as the elderly, is increasingly recognised. Consumers also rarely have any contact with the animals that produce their own food.

In this estranged, efficient world, people struggle to find the moral imperatives to determine the level of welfare that they should afford to animals within their charge. Some, in particular, many companion animal owners aim for what they believe to be the highest levels of welfare provision, while others, deliberately or through ignorance, keep animals in impoverished conditions where their health and well-being can be extremely poor. Today's multiplicity of moral codes for animal care and use are derived from a broad range of cultural influences, including media reports of animal abuse, guidelines on ethical consumption and campaigning and lobbying groups.

This series has been designed to help contribute towards a culture of respect for animals and their welfare by producing academic texts addressing how best to provide for the welfare of the animal species that are managed and cared for by humans. The species-focused books produced so far have not been detailed blueprints for the management of each species, rather they have described and considered the major welfare concerns, often in relation to the wild progenitors of the managed animals. Welfare has been considered in relation to animals' needs, concentrating on nutrition, behaviour, reproduction and the physical and social environment. Economic effects of animal welfare provision were also considered where relevant, as they were key areas where further research is required.

In this volume, we continue the series focus so far of addressing the welfare of one species or a group of species. However, the group of farmed species that are the topic of this book, the ratites, are unusual because they have been farmed for a relatively short period of time, just over 100 years, and are essentially undomesticated. This brings two major problems in comparison with modern farming methods for the traditional species. First, the optimum methods for husbandry of the species in different regions of the world are still in development and, second, the lack of domestication influence and large size of the birds provides further difficulties for husbandry systems, particularly in relation to handling practices. Because of these difficulties, an innovation to the series has been included to consider ethical aspects of the farming of ratites. Pioneering research with ratites to examine their welfare has been undertaken by Dr. Phil Glatz, Senior Research Scientist in Animal Welfare at the South Australian Research and Development Institute. Dr. Glatz, with the support from Dr. Christine Lunam, Senior Lecturer, School of Medicine, Flinders University and Dr. Irek Malecki, Associate Professor,

School of Animal Biology, University of Western Australia, has organised a team of leading scientists experienced in ratite husbandry and welfare to contribute to this volume.

With the growing pace of knowledge in this new area of research, animal welfare science, it is hoped that this series will provide a timely and much-needed set of texts for researchers, lecturers, welfare advocacy groups, policy makers, practitioners and students. My thanks are particularly due to the publishers for their support, and to the authors and editors for their hard work in producing the texts on time and in good order.

St. Lucia, Australia Clive Phillips

References

Food and Agriculture Organisation (2002) http://www.fao.org/ag/aga/index_en.htm

Acknowledgements

The editors are grateful to Mrs. Belinda Rodda, Agriculture Officer, SARDI Livestock and Farming systems for communication with authors and publisher and formatting of references and script.

Contents

Contributors

D. Black ASAEL Consultancy Services, RMB 2235, Moama NSW 2731, Australia, doug@microchips.com.au

T. Brand Department of Agriculture Western Cape, Elsenburg Animal Production Institute, Private Bag X1, Elsenburg, 7607 Stellenbosch, South Africa; Department of Animal Sciences, University of Stellenbosch, Stellenbosch 7600, South Africa, TersB@elsenburg.com

S.W.P. Cloete Department of Animal Sciences, University of Stellenbosch, Matieland 7602, South Africa; Institute for Animal Production, Elsenburg, Private Bag X1, Elsenburg 7607, South Africa, SchalkC@elsenburg.com

R.G. Cooper Flat 5, Eurohouse, Dog Kennel Lane, Walsall WS1 2BU, England, UK, rgcooperuk@yahoo.com

D.C. Deeming Department of Biological Sciences, University of Lincoln, Riseholme Park, Lincoln LN2 2LG, UK, cdeeming@lincoln.ac.uk

P.C. Glatz SARDI, Roseworthy Campus, University of Adelaide, Roseworthy, SA 5371, Australia, phil.glatz@sa.gov.au

L.C. Hoffman Department of Animal Sciences, Stellenbosch University, Private Bag X1, Matieland 7602, South Africa, lch@sun.ac.za

H. Lambrechts Department of Animal Sciences, Stellenbosch University, Private Bag X1, Matieland 7602, South Africa, helet@sun.ac.za

C.A. Lunam Flinders University, GPO Box 2100, Adelaide, SA 5001, Australia, Chris.lunam@flinders.edu.au

I.A. Malecki School for Animal Biology M092, Faculty of Natural and Agricultural Sciences, The University of Western Australia, Crawley, WA 6009, Australia, irek.malecki@uwa.edu.au

M.B. Martella Centro de Zoología Aplicada, Universidad Nacional de Córdoba, Rondeau 798, Córdoba 5000, Argentina, martemo@efn.uncor.edu

J.L. Navarro Universidad Nacional de Córdoba – Consejo Nacional de Investigaciones Científicas y Técnicas, Centro de Zoología Aplicada, CC 122, Córdoba 5000, Argentina, navarroj@efn.uncor.edu

A. Olivier Ostrivet, Kooperasie Straat, Oudtshoorn 6620, South Africa, aolivier@k3sa.com

C.J.C. Phillips Centre for Animal Welfare and Ethics, School of Veterinary Science, University of Queensland, Gatton 4343, QLD, Australia, c.phillips@uq.edu.au

P.K. Rybnik-Trzaskowska Jagiellonska 44/75, 03-462 Warsaw, Poland, prybnik@gmail.com

G. Tulloch Centre for Animal Welfare and Ethics, School of Veterinary Science, University of Queensland, Gatton 4343, QLD, Australia, g.tulloch@griffith.edu.au

K.A. Weir School of Biomedical Sciences, University of Queensland, St Lucia, QLD 4072, Australia

B. Rodda

D. Black

S.W.P. Cloete

R.G. Cooper

D.C. Deeming

P.C. Glatz

L.C. Hoffman

H. Lambrechts

C.A. Lunam

I.A. Malecki

M.B. Martella

J.L. Navarro

A. Olivier

C. Phillips

P.K. Rybnik-Trzaskowska

G. Tulloch

K.A. Weir

Chapter 1
The Ethics of Farming Flightless Birds

G. Tulloch and C.J.C. Phillips

Abstract The ethics, or morality, of farming a relatively novel and undomesticated group of animals, the ratites, is considered. Ethical considerations for animal management centre on their right to life, bodily health and integrity, opportunity to use their senses and emotions, to have affiliations with conspecifics and be part of a worldwide species network, to play and to have control over one's environment. Ratites are considered to present greater ethical problems compared to conventional animal farming because of their inherent unsuitability for farming for meat and other products and their limited level of domestication. This unsuitability arises principally from their large size, slow maturation and limited social structure relative to other farmed birds. The absence of a domestication influence to reduce aggression and flight distance means that they have a significant potential to inflict damage on themselves, their handlers and conspecifics. Bodily mutilations, such as declawing may mitigate damage to others, but is ethically questionable because of potential welfare impact and offence to integrity. It is concluded that significant ethical concerns surround ratite farming that make the practice of dubious value as a means of producing food and leather with due respect to the animals' needs.

Keywords Ethics · Morality · Ostrich · Ratites

1.1 Introduction

The study of animal ethics is concerned with whether our behaviour in relation to animals is morally defensible and correct. Common ethical concerns relating to animals include their welfare, the use to which we put them, artificially reduced longevity, challenges to bodily integrity, genetic modification, the impact of animals on the environment and humans and the use of animals in religious practices

G. Tulloch and C.J.C. Phillips (✉)
Centre for Animal Welfare and Ethics, School of Veterinary Science, University of Queensland, Gatton 4343, QLD, Australia
e-mail: g.tulloch@griffith.edu.au; c.phillips@uq.edu.au

P.C. Glatz et al. (eds.), *The Welfare of Farmed Ratites*, Animal Welfare 11,
DOI 10.1007/978-3-642-19297-5_1, © Springer-Verlag Berlin Heidelberg 2011

(Phillips 2009). Our beliefs about animals are likely to have a direct influence on the way in which we conduct our behaviour towards them. Different stakeholders in the animal industries have different viewpoints and it is important to understand and consider the broad range of perspectives that may be held. For example, in the case of using ratites for food production, the views of producers, transporters, veterinarians and consumers have to be included in any assessment of the morality of the process. Typically people have views based on the utility of the outcomes or their beliefs about our responsibility to animals or some combination of these two factors. Other perspectives incorporate the view that the community in which people live is responsible for determining our actions, or that we develop a contract with animals, based on benefits to both. An understanding of the historical development of the different views may assist in identifying their importance to society.

This chapter explores the ethics of farming ratites – the family of flightless birds that includes emus, ostriches, cassowaries and moas. The first step is to consider the field of animal ethics – what it involves, and significant conceptual developments in its evolution to the present. With this groundwork laid, we will then be in a position to outline an ethical framework against which to assess the issues relating to the farming of ratites.

1.1.1 The Ethics of Human Use of Animals

Animal ethics has not always been seen as a cause for concern.

Animals have long been considered inferior to humans and different in kind, not merely in degree – though this firm boundary was problematised by Darwin's 'The Origin of Species' (Darwin 1859). In Judaeo-Christian ethics, God gave humans dominion over animals – moderated by injunctions towards kindness. The mediaeval notion of the great Chain of Being, with man at the apex, expressed this. The philosopher Kant (1997) argued that animals were not rational or autonomous, and so their lives were not ends in themselves. In Kant's view, presented in 'Lectures on Ethics', our duties to animals are merely indirect duties towards humanity, and if we treat animals kindly, we strengthen the disposition to behave kindly towards humans – like exercising a moral muscle on a proxy object. The corollary for Kant was that animals could appropriately be treated as means to our ends. For Kant, moral duties can only be to self-conscious beings. Only such beings can be members of the moral community. Animals could thus be relegated to beings of secondary concern – if concern at all – for want of a soul, of rationality (construed in a particular, narrow way), of autonomy or of language.

The Christian notion was, at best, one of human stewardship and at worst, human dominion over the rest of nature, including animals. This exacerbated the long-established prejudice in western culture in favour of rationality as the defining and unique characteristic of human beings.

In the Enlightenment, Descartes (1901) argued that like clocks or robots, animals were but machines that moved and made sounds but had no feelings. In such a context, it was easy to portray animals as quasi-clockwork animated robots – 'furry clocks'. Such a conception rationalised vivisection, for creatures with no consciousness could feel no pain.

1.1.2 *Sentience*

Jeremy Bentham, the founder of utilitarianism, was the first major figure in Western ethics to advocate in 1789 the direct inclusion of animals in our ethical thinking. As he memorably argued:

What else is it that should trace the insuperable line? Is it the faculty of reason or perhaps the faculty of discourse? But a full-grown horse or dog is beyond comparison a more rational, as well as a more conversable animal than an infant of a day or a week, or even a month old. But suppose they were otherwise, what would it avail? The question is not Can they reason? nor Can they talk? But Can they suffer?

In this way, Bentham (1789) addressed the issue of the boundary between human and animal and introduced the concept of sentience – or the capacity to feel pleasure and pain – as the central criterion of issues of animal ethics. This was the driving force behind the POCTA – prevention of cruelty to animals – tradition of legislation, which still prevails today. It is an animal welfare framework, evident in the RSPCA and the work of some animal activists.

Singer's (1990) work is grounded in this Benthamite tradition, and he further argues that the difference between humans and animals is one of degree, not of kind, i.e. not absolute, and that the boundary is quite porous.

1.1.3 *Circles of Compassion*

As early as the second century AD, the Stoic philosopher Hierocles created a vivid metaphor for extending the boundaries of our moral concern. Imagine, he argued, that each of us lives in a series of concentric circles, the nearest being our own body, and the furthest being the entire universe. The task of moral development is to move the outer circles progressively to the centre, so that one's relatives become like oneself, strangers like relatives, and so on. Singer (1990) adopts this metaphor, and argues for explicitly extending the circle of one's concern beyond the boundary of one's own species, to include animals, and, ultimately further, to the whole environment. Why we should do this is meant to be intuitively obvious; at least, learning to see it so is the path of enlightenment in some religions. Humans appear to have built-in resistance, however.

1.1.4 *Speciesism*

Speciesism was the second great driving idea in animal ethics after sentience. It was a term coined by Ryder in 1970s (Ryder 2005) and popularised by Singer (1990).

It means a prejudice or attitude of bias in favour of members of one's own species against those of members of another species. Speciesism obviously picks up on the unfavourable connotations of racism and sexism, and the movements to extend equal consideration to the interests of coloured people and of women.

The task to change deep-seated, unreflective notions of the species barrier is the task we face now, and it is perhaps the hardest of all, because the attitudes are so entrenched, and the economic incentives to persist with cost-cutting, production-line, inhumane treatment of animals are so great. Pope Benedict (2005) has condemned the 'industrial use of creatures, so that geese are fed in such a way as to produce as large a liver as possible, or hens live so packed together that they become just caricatures of birds'. It is in this context that the argument to expand our circle of compassion appeals to considerations of animal welfare, but also makes a transition to animal rights, as sentient beings who deserve quality of life. There may be a common perception that birds are less worthy of high standards of animal welfare than mammals, in part because we empathise more easily with the latter. Birds are rated by humans as less sentient than mammals, but more sentient than fish (Phillips and McCulloch 2005; Meng 2009; Meng et al. 2009), although there is no physiological evidence for the validity of these differences. Certainly the concept of sentience is central to attributing animals' welfare considerations, as is an opposition to cruelty, which is its corollary. But the focus of concern for many animals is primarily negative, with an indirect appeal to empathetic identification only for those animals most like us. Appealing to quality of life – whether human or animal – needs specification if it is to be more than vague.

There now seems to be an even better theoretical approach, which is more broad-ranging and specific, and grounds positive guidance for action. It is the capabilities approach, advocated by Nussbaum and Sen (1993), the latter a Nobel prize-winning economist, who pioneered a Quality of Life approach to human capabilities in the context of aid and human development, tied to the UN Declaration of Human Rights.

1.1.5 The Capabilities Approach as an Ethical Framework

The capabilities approach was first articulated in 'The Quality of Life' (Nussbaum and Sen 1993), based on their research in a World Institute for Development Economics Research (WIDER) study for the U.N. University. The book comprises papers from a 1988 Conference in Helsinki, which they organised for WIDER, where Nussbaum spent a month in the summer for 8 years in residence. Till then she had thought little about problems of global justice or feminist philosophy. Her time there transformed her work.

Aristotle's insistence on the importance of individual perception of concrete circumstances, she felt, had a contribution to make to a field that is 'frequently so pre-occupied with formal modelling and abstract theorising that it fails to come to grips with the daily reality of poor people's lives'.

WIDER's mandate is to engage in interdisciplinary research, and the 1988 conference brought together economists and philosophers around the question what is meant by 'quality of life' and what is required in terms of social policy for improving it'.

A crude measure of per capita income is generally taken as indicative of human welfare, which begs important questions such as the distribution of wealth and income, and the need to assess a number of distinct areas of human life. At the micro level, the notion of maximising an individual's utility underlies much of conventional demand theory. But this raises two questions: is utility measurable, and is it the right thing to be measuring when we are interested in assessing the quality of human lives? Nussbaum and Sen (1993) suggest we should instead measure people's capabilities, what they are able to do and to be in a variety of areas of life.

The ten capabilities listed ranged over several areas:

1. Life
2. Bodily health
3. Bodily integrity
4. Senses, imagination and thought
5. Emotions
6. Practical reason
7. Affiliation
8. Other species
9. Play
10. Control over one's environment

The influence of this approach is shown by the fact that since 1990, Human Development Reports of the UN Development Program have looked at capabilities. Nussbaum was critical of the per capita gross national product interpretation of Quality of Life on two grounds: it does not address distribution or different, non-economic aspects of human life.

In the field of animal ethics, the capabilities approach, as extended by Nussbaum and Sen (1993), appeals for animal welfare based on rights derived from their capabilities – which are outlined. The approach lists ten capabilities, nine of which also apply to animals. It stresses how much more has to be considered and provided for than is implied by sentience, and covers the whole range of animals, including in zoos, rodeos, museums and laboratories. It involves a radical paradigm shift in outlook, and has huge practical implications. It's observable, and it's easy to identify where the shortcomings fall. This makes it both the most current and the most exciting development in animal ethics.

Let us now examine in detail the capabilities, as applied to animals.

The first is *Life*, which entails animals are entitled to continue their life, whether or not they take a conscious interest in it. This puts pressure on the meat industry to reform its practices, as well as problematising killing for sport (hunting and fishing) and for fur.

Bodily health is the second entitlement, and where animals are under human control, this entails laws banning cruel treatment and neglect, confinement and ill treatment of animals in meat and fur industries; forbidding harsh or cruel treatment for working animals, including circus animals, and regulating zoos, aquaria and parks, as well as mandating adequate nutrition and space. Nussbaum and Sen (1993) point to the anomaly that animals in the food industry are not protected as domestic animals are, and recommends that this anomaly be eliminated.

Bodily integrity is the third entitlement, which would prevent the declawing of ostriches (Meyer et al. 2002) and other mutilations, such as tail-docking, that make the animal more beautiful to humans. It would not ban forms of training that are part of the characteristic capability profile, such as training horses or border collies.

Senses, *imagination*, and *thought* constitute entitlement 4, and entail access to sources of pleasure such as free movement in an environment to please the senses, and which offers a range of characteristic activities.

Emotions are entitlement 5. Nussbaum and Sen (1993) argue that all animals experience fear, and many experience anger, resentment, gratitude, grief, envy and joy, while a small number can experience compassion. Hence they are entitled to lives where it is open to them to have attachments to others, and not have these attachments warped by isolation or fear. While this is understandable in relation to domestic animals, it is overlooked in relation to zoo and farm animals and research animals.

Practical reason (entitlement 6) is 'a key architectonic entitlement in the case of human beings' and has 'no precise analogues in the case of non-human animals'. However, we should consider the extent to which the being has a capacity to frame goals, and support it if this is present, as well as providing plenty of opportunity for movement and variety of activities.

Affiliation is entitlement 7 on the capabilities list. Nussbaum and Sen (1993) argue that animals are entitled to form attachments, and to relations with humans that are rewarding rather than tyrannical, as well as to live in 'a world public culture that respects them and treats them as dignified beings'.

Other species is capability 8, and calls for the formation of an 'interdependent world in which all species will enjoy cooperation and mutually supportive relations with one another'. This idealistic entitlement calls, in Nussbaum and Sen's (1993) words, 'for the gradual supplementation of the natural by the just'.

Play is capability 9, and is central to the lives of all sentient animals. It entails adequate space, light and sensory stimulation and the presence of members of other species.

Control over one's environment is capability 10, and has two aspects in the case of humans – political and natural. For non-human animals, it entails being respected and treated justly, even if a human guardian must go to court, as with children, to vindicate those entitlements. The analogue of human property rights is respect for the territorial integrity of their habitat, domestic or wild, and the analogue of work rights is the rights of labouring animals to dignified and respectful labour conditions.

Only Practical Reason does not fit smoothly with animals, and much of what it requires can be derived from the criteria for flourishing. However, even excluding it, if the other nine of these ten capabilities were taken seriously, it would transform the common conception of how much needs to be provided as basic conditions for animals – not just life, health, and the maintenance of bodily integrity, but opportunities to experience the senses, imagination and thought, emotions, affiliation, relations with other species, play and control over the animal's environment. Yet it is hard to think of a single instance where these capabilities are currently allowed for.

Nussbaum and Sen (1993) recognise that these rights need international cooperation, via accords, such as the U.N. Declaration of Human Rights, as well as the ineliminability of conflict between human and animal interests. Some bad treatment of animals, she argues, can be eliminated without serious loss of human well-being. In the use of animals for food, for example, she suggests setting the threshold on focussing on good treatment during life and painless killing. In the use of animals for research, she argues much can be done to improve the lives of research animals, without stopping useful research. It is unnecessary and unacceptable for primates used in research to live in squalid and lonely conditions. Nussbaum and Sen (1993) advocate asking whether the research is really necessary; focussing on the use of less complexly sentient animals; improving the conditions of research animals including terminal palliative care; removing psychological brutality; choosing topics cautiously so no animal is harmed for a frivolous reason; and making a constant effort to develop experimental methods (such as computer simulation) that do not have bad consequences. The three Rs Replacement, Refinement and Reduction first espoused by Russell and Burch (1959) – has some affinity to Nussbaum and Sen's (1993) approach here.

Phillips (2009) recently suggested expanding the basis for an ethical framework to include the genetic integrity of animals, focussing on our duty towards animal species. He assesses our interactions with animals under the following concerns: their welfare, their ability to display choices, the use to which we put them, our impact on their longevity, challenges to their bodily and genetic integrity and the impact of animals on the environment and humans. These issues have been used as a basis for surveying attitudes to animals and indices developed to investigate these issues in different cultures (Meng et al. 2009). Contrary to this extension of our responsibilities, Roger Scruton has suggested that we should reduce our responsibilities so that, for wild animals at least, our principle duty is to animal species, not individual animals (Scruton 1996). Thus Scruton is able to justify hunting animals because individual animals are not worthy of our consideration, only species.

1.1.6 The Capabilities Approach to Ratites

Adopting the capabilities approach to the ethics of farming ratites, it is clear that the first priority is to learn about their nature and needs. At the moment, there is a glaring deficiency here, as very little is known about their needs, and such research done tends to focus on a small range of fairly obvious issues: that they are an endangered species; that their habitat is at risk; that humans feed them and by thus interfering with them, assume a responsibility for them. There is an apparent symbolic association with masculinity and violence (Nihill 2002), and – perhaps of primary interest – they present an attractive potential for farming, as they are able to exist in a broad range of climatic conditions, although not without welfare risks as is explained later. Their meat is valued for its nutritional content: low in cholesterol, high in protein and of low fat content while their skin, feathers and oil are used widely.

1.1.7 Ethical Issues in Farming Ratites

The animals we farm for food and other products are principally species that have been domesticated to make them more amenable to the farming process. The most efficient animals to produce meat are those in the early growth stages, before growth declines and the maintenance cost of the animals assumes a significant cost. Thus animals killed for meat are usually slaughtered at approximately 50% of their mature size. To efficiently produce large numbers of offspring capable of growing rapidly to this stage the species used are naturally prolific (to minimise the number of breeding females), polygynous (to minimise the number of males in the breeding herd/flock), early maturing (minimising the cost of rearing replacement breeders), gregarious (reducing the tendency for animals to stray and allowing the animals to be herded) and herbivorous (to invoke a direct transfer of plant energy to meat energy, without the inefficiencies of passing through another process in the food chain by farming carnivores). Ratites can only be claimed to possess two of these virtues that would make them suitable for farming for food, their mainly herbivorous diet and prolificacy. They can produce many young each year, up to 100, which can be artificially reared although as noted below this raises ethical issues. They are not early maturing, with ostrich hens starting to produce eggs at about 18–36 months. They are usually bred in pairs or trios and have small social groups. Moreover, they have not been domesticated, which is the process by which wild animals are tamed to allow them to be kept more easily in intensive farming systems. Domestication allows animals to tolerate the presence of humans more readily, reduces aggression and often reduces their size so that they are more easily handled. Typically large numbers of farm animals, such as cattle or sheep, can be moved by one or two humans, perhaps with the aid of a dog. This is not possible with ratites that require very careful handling (see Chap. 10) and is prone to stress-related disorders during and after transport (Kamau et al. 2002). In addition, ostriches stand at up to eight feet tall, making them potentially dangerous animals to handle.

Ratites have not been domesticated and are naturally very aggressive (especially in the breeding season) in their relationships with humans. Chicks can be imprinted on humans, lessening their intuitive aggression towards them, but in adulthood revert to wild type and show clear evidence of aggression (Nihill 2002). Ostriches and cassowaries are the only birds that have killed humans by physical attack, and there have been many incidents of serious injury when humans have attempted to feed cassowaries or hold them in captivity (Kofron 1999). Ostriches, like other ratites, are large compared with other farm animals, often in excess of 2 m. They are difficult to handle, often running if they are stressed, running into fences, running until exhausted (Hoffman and Lambrechts 2011). They need space to run, usually several acres, which are often not provided in intensive, feedlot-type operations. The birds are easily frightened by novel stimuli. Capture myopathy (see Chap. 11), similar to that experienced in captured wild animals, accounts for some of the serious mortality that can eventuate following transportation (Hoffman and Lambrechts 2011; Navarro and Martella 2011).

It is clear that standards for transport, feeding, intensity of housing are often not sufficiently supported by scientific research and are based primarily on expert opinion. Transport is a particularly stressful period for the birds and it should be a pre-requisite for new species farming that welfare standards are adequately evaluated *before* initiation of the practice. Some of the knowledge gained from other farmed species will benefit ratites. For example, they are now known to suffer from the same depletion of glycogen reserves, high pH and consequent dark muscle when stressed at slaughter that cattle are prone to (see Chap. 10). The research required to optimise the ratite farming systems will be less than has been conducted with cattle and sheep in the twentieth century, but still substantial in relation to the size of the industry.

The small size of the industry and difficulties in managing ratites in farming systems means there are not many skilled stockpeople that can care for the animals in new enterprises. In the absence of indigenous, inherited knowledge, training of all stockpeople should be compulsorily undertaken (see Chaps. 5, 6 and 9). This book is therefore an attempt to summarise the current state of knowledge, but it also points out gaps in the literature that need filling.

Another ethical concern relates to the removal of eggs soon after lay for artificial incubation, allowing the hen to return to lay again more rapidly (see Chaps. 4 and 11). This practice is commonplace, but the incubation is often not successful (Deeming 2011). This is comparable to calf removal and artificial rearing in the dairy industry, except that the mortality rate in this case is much lower. Still this practice has been the cause of ethical concern on account of its unnaturalness and threats to the survival of the birds unless considerable experience has been gained. Deeming (2011) raises doubt about the sentience capacity of birds in ovo and hence their capacity to suffer. However, even if unable to suffer, threats to the survival of the bird challenges other ethical values, in particular the right to life of the embryo.

Other ethical concerns include the ‘assistance’ given to birds during hatching, which can jeopardise their future survival and welfare (see Chap. 4), Nussbaum and Sen’s (1993) first capability. Periods of human interference with the birds, e.g. during transport, have a much greater effect than with domesticated poultry. This raises the ethical question as to why such birds are kept for meat, leather, feather and oil production if they are more difficult to keep in a high welfare state than other, more efficient birds. One reason is to satisfy some humans’ desire for variety in the diet, another is that it potentially allows them to claim dominance over a greater number of species.

Ratites are often kept in farms with very different climates to their native habitat. There is particular concern about keeping ostriches in northern Europe, where winter temperatures affect the birds' behaviour and require a significant increase in food consumption and provision of shelter if the birds' thermoregulatory needs are to be met (Deeming 1998).

Leather is one of the major products from ostrich farming, but its value is often reduced by scratches and kick marks produced during interbird aggression, handling and interactions between chicks when huddling (see Chaps. 4 and 6). Declawing provides a potential solution but can be criticised on the grounds that bodily integrity is compromised. It is commonly practised with emus both to reduce damage to the leather and also risk to handlers. Although relatively unknown in ostriches, if proved successful it could still result in chronic pain and/or loss of locomotive ability or other behavioural capabilities. It needs further study, both of potentially adverse effects on behaviour immediately after the practice, and long-term adverse effects.

1.1.8 Conclusion and Research Needs

Whereas a study of animal welfare is able to be conducted through established scientific procedures, research in animal ethics is focused more on evaluating attitudes, their origin and their impact on the process, for example the impact that consumer attitudes have on purchasing behaviour, or the effect of producer attitudes to painful procedures on their performance of husbandry practices. Understanding differences in attitudes and the viewpoints of different cultures will help to foster a tolerant and equitable attitude, and is especially important in multicultural societies and in relation to international trade in animals and products derived from them. Little is known about the attitudes to ratite farming, particularly in relation to other forms of animal farming. Ratites are relatively new to farming and the industry has experienced rapid growth without a sound foundation of research and gradual learning, such as was developed for the livestock industry. This rapid development brings risks of unethical practices, and because the numbers used are still relatively small the development of Codes of Practice and standards has been slow and limits the regulatory control of practices.

References

Bentham J (1789) Principles of morals and legislation. Clarendon, London

Darwin C (1859) On the origin of species by means of natural selection, or the preservation of favoured races in the struggle for life. John Murray, London

Deeming D (1998) A note on effects of gender and time of day on the winter time-activity budget of adult ostriches (Struthio camelus) in a farming environment. Appl Anim Behav Sci 59: 363–371

Deeming DC (2011) Incubation and chick rearing. In: Glatz P (ed) The welfare of farmed ratites. Springer, Heidelberg

Descartes R (1901) Meditations on first philosophy (first published in Latin and French in 1641). Cambridge University Press, Cambridge, Edited John Veitch

Hoffman LC, Lambrechts H (2011) Bird handling, transportation, lairage and slaughter: implications for bird welfare and meat quality. In: Glatz P (ed) The welfare of farmed ratites. Springer, Heidelberg

Kamau JM, Patrick BT, Mushi EZ (2002) The effects of mixing and translocating juvenile ostrich chicks (Struthio camelus) in Botswana on the heterophil to lymphocyte ratio. Trop Anim Heal Prod 34:249–256

Kant I (1997) The Cambridge edition of the works of Immanuel Kant. Cambridge University Press, Cambridge

Kofron CP (1999) Attacks to humans and domestic animals by the southern cassowary (*Casuarius casuarius johnsonii*) in Queensland. Aust J Zoo 249:375–381

Meng J (2009) 'Origins of Attitudes of Animals'. Google book online http://jmeng.goodeasy.info/publications/readOAA.php, accessed 5 April

Meng J, Hao LP, Hou H, Illmannová G, Alonso ME, Hanlon A, Aldavood SJ, Choe BI, Lee GL, Handziska A, Kjastad H, Lund V, Olsson A, Rehn T, Keeling LJ, Pelagic VR, Kennedy M, Phillips CJC (2009) Attitudes to animals in Eurasia: the identification of different types of animal protection through an international survey. Abstract number IS OP061, In: Proceedings of the Minding Animals Conference, Newcastle, July

Meyer A, Cloete SWP, Brown CR, van Schalkwyk SJ (2002) Declawing ostrich (Struthio camelus domesticus) chicks to minimize skin damage during rearing. S Afr J Anim Sci 32:192

Navarro JL, Martella MB (2011) Ratite conservation: linking captive-release and welfare. In: Glatz P (ed) The welfare of farmed ratites. Springer, Heidelberg

Nihill N (2002) Dangerous visions, the cassowary as good to think and good to remember amongst the Agnanen. Oceania 72:258–274

Nussbaum M, Sen A (1993) The quality of life. Clarendon, London

Phillips CJC (2009) The welfare of animals: the silent majority. Springer, Dordrecht

Phillips CJC, McCulloch S (2005) Attitudes of students of different nationalities towards animal sentience and the use of animals in society, with implications for animal use in education. J Bio Educ 40:17–24

Pope Benedict XV1 (2005) How Pope Benedict XVI views animals. www.all-creatures.org/living/howpope.html. Retrieved 19 April

Russell WMS, Burch RL (1959) The principles of humane experimental technique. Methuen, London

Ryder R (2005) "All beings that feel pain deserve human rights". The Guardian. 6 August. http://www.guardian.co.uk/animalrights/story/0,11917,1543799,00.html. Retrieved 19 April, 2010

Scruton R (1996) Animal rights and wrongs. Demos, London

Singer P (1990) Animal liberation: a new ethics for our treatment of animals, 2nd edn. Random House, New York

Chapter 2
Breeder Welfare: The Past, Present and Future

S.W.P. Cloete and I.A. Malecki

Abstract The welfare needs of mature ratite breeders are reviewed in this chapter. Ratite reproductive strategies and mate choice are discussed with reference to compatibility of breeding males and females housed in pairs. Past and present mating structures are discussed in terms of behaviour needs. It was shown that trauma is associated with the majority of cases where mature pair-bred ostrich breeding birds exit the breeding flock prematurely. It was noted that small (< 20 breeding birds) breeding colonies probably approached the mating system prevalent under natural conditions closest, and that such systems may be preferred if birds are confined to small areas. Larger colonies of up to 200 breeding birds may be kept, provided that the areas used are sufficiently large for the birds to disperse and form natural breeding groups. Preliminary results suggest that male aggression in ostriches is heritable. Measures of temperament that allow breeding ratites to adapt to farmed conditions should urgently be identified. Genetic and environmental (co) variances of such traits should be estimated. Conventional quantitative genetics methods should subsequently be used to improve such traits, if feasible. The present knowledge of the genomes of farmed ratites should be expanded simultaneously, should appropriate funding be forthcoming. There appears to be ample scope for the improvement of welfare in farmed ratite breeders.

Keywords Colony · Compatibility · Pair · Reproductive behaviour · Sustainable

---

S.W.P. Cloete (✉)
Department of Animal Sciences, University of Stellenbosch, Matieland 7602, South Africa
and
Institute for Animal Production, Elsenburg, Private Bag X1, Elsenburg 7607, South Africa
e-mail: SchalkC@elsenburg.com

I.A. Malecki
School for Animal Biology M092, Faculty of Natural and Agricultural Sciences, The University of Western Australia, 35 Stirling Highway, Crawley, WA 6009, Australia
e-mail: irek.malecki@uwa.edu.au

P.C. Glatz et al. (eds.), *The Welfare of Farmed Ratites*, Animal Welfare 11,
DOI 10.1007/978-3-642-19297-5_2, © Springer-Verlag Berlin Heidelberg 2011

2.1 Background

It is generally accepted that commercial ostrich farming started in South Africa about one and a half centuries ago (Smit 1964; Osterhoff 1979). The commencement of farming with other ratites (emus and rheas) has been much more recent. This implies that farmed ratites have had a much shorter period to adapt to the farming environment and farming routines than the other major livestock species, which have been domesticated for centuries.

Problems like a relatively low and variable egg production in ostriches (Bunter 2002; Cloete et al. 2008b) and a low survival rate of rhea (Labaque et al. 1998; Barri et al. 2005) and ostrich (Cloete et al. 2001) chicks have been attributed to the failure to adapt to the farming environment. This review seeks to:

- Assess welfare in the context of the various ratite species
- Discuss reproductive behaviour of ratites from a welfare perspective
- Investigate production systems for maintaining maximum output from reproducing ratite breeder birds, while also meeting their welfare needs
- Discuss feather harvesting from mature breeders in the context of welfare
- Assess the impact of the present production systems upon the natural resource
- Discuss priorities for future research involving breeder welfare

Explicit data are scant on the impact of farming routines on the welfare of these birds (Mitchell 1999). However, breeder welfare will be assessed within the context of the commercial production environment drawing on limited scientific studies where breeder welfare was targeted specifically.

2.2 Welfare of Ratites

Welfare of commercial animals and subjects of animal experimentation are usually expected to depend on five freedoms (Mitchell 1999), namely:

- Freedom from hunger and thirst
- Freedom from discomfort
- Freedom from pain, injury and disease
- Freedom to express normal behaviour
- Freedom from fear and stress

Some of these aspects are covered in more detail elsewhere, for example, Natural Mating and Artificial Insemination in Chap. 3, Nutrition in Chap. 4 and Veterinary Topics in Chap. 8. Aspects such as the expression of normal reproductive behaviour, as well as trauma and discomfort of farmed ratites that are caused by injury and a failure to adapt to the commercial environment, are topics that are discussed in this chapter.

In general, behaviour repertoires can be divided into normal behaviour patterns and aberrant behaviour routines, which in general are undesirable.

Normal activities such as twirling, thermoregulatory behaviour, pecking in an exploratory sense, grooming and trembling were observed in farmed Canadian ostriches (Samson 1996). Normal social behaviour patterns such as assuming a threatening posture, kicking, vocalisation and submission were also recorded, while clucking, fluttering and kantling were listed as normal sexual behaviour patterns. On the other hand, Samson (1996) also listed undesirable abnormal behaviour patterns like feather-pecking, toe and face pecking, behavioural star-gazing, anorexia, adipsia, dietary indiscretion, pica and overt aggression in the same birds, while Bubier et al. (1998) regarded kantling as undesirable when performed to a human. The ostriches studied were more prone to aberrant behaviour during periods of confinement. With the exception of feather pecking and an increase in homosexuality under high concentrations of breeding ostriches in open camps (Lambrechts et al. 2004), such aberrant behaviour patterns are not common in mature breeding ostriches under normal production regimes in South Africa. Owing to a lack of sheltering behaviour in even mature breeding ostriches (Deeming and Bubier 1999), they can compromise their welfare by sitting in the open and freezing to death in countries with extreme winter climates even when adequate shelter structures are available.

2.3 Ratite Reproductive Behaviour

Out of the breeding season, free-ranging ostriches, emus and rheas are gregarious, and form groups of birds of mixed age without an apparent social structure (Deeming and Bubier 1999). In contrast, cassowaries and kiwis are considered to be more solitary out of the reproductive phase (Ridley 1978). However, it is important to consider the different breeding strategies employed by the respective ratite genera, as it has an important impact upon their welfare under commercial conditions. It is also interesting to note that incubation and care of the newly hatched chicks are provided by the male parent in the majority of ratites (Ridley 1978).

Breeding of commercial ratite stock is also complicated by preferences for specific mates, as reported for ostriches by Deeming (1996) and experienced in emus (I.A. Malecki unpublished). Apart from an obvious loss of reproductive potential, such preferences could also seriously compromise welfare due to isolation, loss of preferred mates and an increased risk of attack and fighting, particularly when they are housed in relatively small areas. It is conceivable that the loss of a preferred mate would be associated with a level of stress and a lack of freedom of expressing normal sexual behaviour in both genders, while aggression prompted by mate incompatibility could seriously compromise welfare by inciting fear, stress and discomfort as well as by inflicting serious bodily harm that could result in permanent incapacitation and death. Against this background, it is also important to review the limited information that is available on mating preferences in ratite species.

2.3.1 *Mating Systems and Reproductive Strategies*

Sexual behaviour, reproductive anatomy and the physiological process of reproduction are adequately described for most members of the ratite family (Deeming and Bubier 1999; Soley and Groenewald 1999; Sales 2006, 2007; Deeming 2009) and will not be reviewed explicitly. However, social structures during the reproductive phase have an important bearing on the commercial mating systems and they are reviewed here. More detailed attention will thus be focused on social structures and reproductive strategies of free-ranging birds. In most cases, these strategies were subjected to previous reviews and will therefore be discussed briefly.

2.3.1.1 Ostrich (*Struthio camelus* spp.)

This group of ratites employ a promiscuous mating strategy, in which both males and females have multiple sexual partners. The mating strategy involves a dominant male that attracts females to his territory. He then mates with these females, which lay eggs in a communal nest. According to the review by Deeming and Bubier (1999), the breeding harem usually consists of a dominant female and one to five subordinate females (three on average). The colony structure and its dynamic may vary depending on the colony size, available area and environmental factors such as food availability or predation. In a study conducted in Australia (Ledger and Malecki 2008) on a colony of 125 birds (90 females and 35 males) fenced on 13.74 ha of land, it has been shown that some of the most productive nests had a rate of egg accumulation of 8.4 per day, suggesting that at least 16 females would be associated with such a nest. Most nests have an accumulation rate of 1.7 eggs per day, suggesting 3–4 females associated with such nests (Ledger and Malecki 2008). Typical numbers of eggs in a nest were reported as 26, with a range of 25–39 in the review of Deeming and Bubier (1999). The number of eggs in a nest accordingly ranged from 27 to 36 in the subsequent study of Kimwele and Graves (2003). Dominant females were reported to contribute ~11 eggs (range 9–14) to the nest (Deeming and Bubier 1999). Based on nest demographics, eggs can be defined as central eggs, which receive better parental care (and thus have a higher probability of hatching) than peripheral eggs. The number of central eggs ranged from 19 to 27 in the study by Kimwele and Graves (2003), while peripheral eggs ranged from 3 to 18. Fertile central eggs produced by the dominant female averaged 6.75 (range from 5 to 8), while 4–12 fertile eggs were produced by subordinate females (Kimwele and Graves 2003). Of these fertile eggs, respectively 2–7 and 2–4 were fertilised by the dominant male in the central and peripheral compartments. The dominant male shares incubation duties with a dominant female when the incubation of the nest commences (Deeming and Bubier 1999). In trios (one male and two females) a second female was reported to share incubation duties, with females usually incubating the eggs during daytime and the male at night. Upon hatching of the chicks after approximately 42 days of incubation, parental care of the brood is undertaken by the dominant male and dominant female (Deeming and Bubier 1999; Deeming 1996).

Dominant females in specific nests also commonly act as subordinate females in the nests of other territorial males (Deeming and Bubier 1999; Kimwele and Graves 2003). The promiscuous sexual strategy of the ostrich is perhaps best illustrated by the parentage structure in two commercial colonies in South Africa (Fig. 2.1a, b;

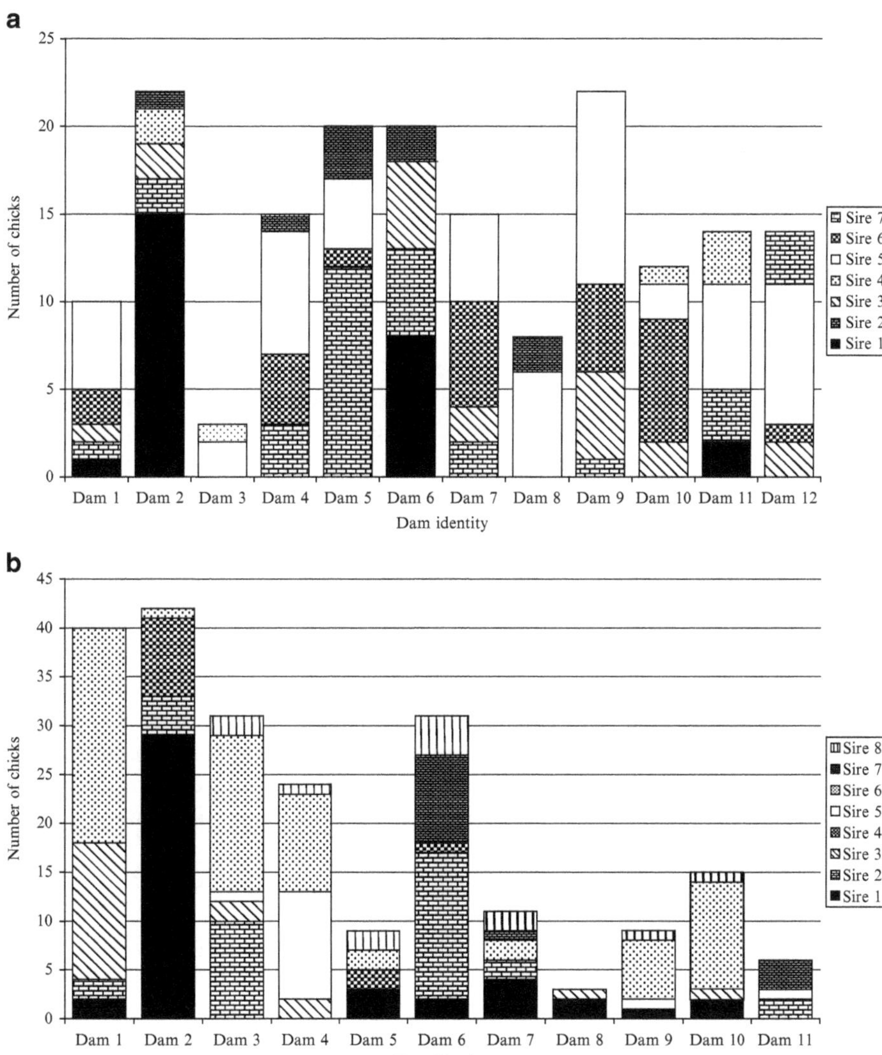

Fig. 2.1 The number of chicks produced by ostrich females in combination with the available males in Colony 1 (**a**) and Colony 2 (**b**). Colony 1 consisted of 12 females and 7 males, and Colony 2 of 11 females and 8 males.
Source: Bonato (2008)

Bonato 2008). It is evident that large differences occurred among the number of chicks produced by different female colony members. In Colony 1, chicks produced by individual females ranged from 3 to 22 chicks. The corresponding range in Colony 2 was between 3 and 42 chicks. It is also evident that all females produced chicks from at least two sexual partners during the mating season.

2.3.1.2 Greater Rhea (*Rhea Americana*) and Lesser Rhea (*Pterocnemia pennata*)

The mating strategy of rheas closely corresponds with that of the ostrich, males attracting females to lay in a nest they scraped. The size of the harem so formed may be up to 15 females in the greater rhea (Smith 1977). Females will typically lay on alternative days, and ultimate nest size may include approximately 50 eggs. The number of eggs in greater rhea nests ranged from 8 to 56 (Fernández and Reboredo 1998). However, the majority of nests contained 20–30 eggs. The number of eggs per nest averaged 28 and resulted in an average of 14 chicks hatched per nest (Fernández and Reboredo 2007). Hatching success seemed to increase with nest size, but there was a suggestion of lower levels of hatching success in nests containing more than the median number of eggs. This finding was related to a higher level of contamination in larger nests (Fernández and Reboredo 2007). Overall egg production per female tends to be higher in captive greater rhea females (40) than in captive lesser rhea females (18) (Navarro and Martella 2002). However, in contrast with ostriches, the male starts to incubate the eggs a few days after the first egg has been laid and chases away all the females from the communal nest soon after (Smith 1977). Males then undertake the sole responsibility for incubating the eggs for ~39 days (Deeming 2009). Upon the hatching of the brood, the male also solely provides parental care to the hatchlings (Sales 2006; Navarro and Martella 2002). Males typically also attract chicks from other broods for the provision of paternal care to the group of chicks under their care (Codenotti and Álvarez 1998). Adoption of chicks was observed in 23% of adult males caring for broods of their own, and adopted chicks composed 37% of the post-adoption brood. Males adopting chicks showed improved parental care, as reflected by an increased vigilance, more frequent sheltering of chicks and greater aggression towards potential intruders (Codenotti and Álvarez 1998).

When leaving the territory of the first male, females start to consort with other males and produce eggs for their nests. Females may mate with and produce eggs fathered by up to seven males in a single reproductive season (Smith 1977).

Nest site selection by greater rheas was studied by Fernández and Reboredo (2002). Nest sites were not selected randomly, but had a large percentage of shrub cover and a low overall visibility (greater concealment). However, microhabitat could not conclusively be related to rate of egg loss from the nest or incubation failure (Fernández and Reboredo 2002). It was suggested that rheas experience reduced benefits from selecting concealed nest sites at present, possibly because of habitat alteration and the type of predation experienced.

2.3.1.3 Emu (*Dromaius novaehollandiae*)

The emu mating system is based on sequential polyandry, although monogamy and promiscuity have also been observed (Coddington and Cockburn 1995; Blache et al. 2000; Davies 2002; Sales 2007). At the commencement of the laying season, the female initiates at least two nests even if the eggs are not regularly collected (I.A. Malecki and P. O'Malley unpublished). While the reason for such behaviour is unclear it may provide a signal for males to act on those nests. Emu males can also mate with a few females (1–3) (Ridley 1978). Each female produces eggs at approximately 3-day intervals. The male starts incubating when the nest size reaches 10–20 eggs, although desire to incubate is highly variable with some males commencing incubation with only 2–4 eggs in the nest (Blache et al. 2000). Even objects similar in shape and colour to the emu eggs can drive incubation behaviour (Blache et al. 2000). The female producing the eggs wanders off to consort with other males. In a pair mating structure, it has been observed that a female would produce on average 3.4 clutches of eggs of 6.7 eggs per clutch over ~84 days (Sales 2007). The eggs are incubated for ~56 days and the hatchlings are provided with parental care solely by the male for more than a year (Sales 2007). The degree of sexual interactions in captive emus can vary with space and stocking density and it may result in high proportion of offspring fertilised by extra-pair copulations. As a result, a high proportion of eggs the male incubates may be fertilised by rival males (Taylor et al. 2000).

2.3.1.4 Cassowary (*Casuarius* spp.)

These birds are mostly solitary, each maintaining their own territory (Ridley 1978). During the mating season they are seen in pairs. The clutch size is 2–5 eggs, while 80% of males are breeding once in 3 years (Moore 2007). The clutch size suggests monogamy (Ridley 1978). Crome (1976) observed a female mating with two males in a single season, but such occurrences do not appear to be common. Incubation and subsequent parental care are provided solely by the male, the chicks remaining with him for 3–4 months post-hatching (Ridley 1978).

2.3.1.5 Kiwi (*Apterix* spp.)

The kiwi female is reported to be bigger and more assertive than the male (on average 2.4 vs. 1.9 kg), but there are reports of fights between males in the wild (Reid and Williams 1975). Brown kiwis live as bonded pairs in fixed territories throughout the year, both members defending the territory during the breeding season (Bassett et al. 2005). Copulation starts about 18–22 days before egg-laying. Kiwi females produce 1, 2 or rarely 3 eggs in a season, and the mating system appears to be monogamy (Reid and Williams 1975; Bassett et al. 2005).

In contrast to other ratites, kiwis dig a carefully concealed nesting burrow, which is ~65 cm long with a nesting chamber at the end (Bassett et al. 2005). It is well known that the kiwi egg is the largest of all bird species relative to female body size (Ridley 1978; Bassett et al. 2005). The male incubates the eggs for 71–84 days. Chicks hatch within 5–13 days from each other and are brooded by the male initially. The male do not feed them in this period. Chicks are precocial and fully feathered with open eyes at hatching, but with a hugely distended abdomen because of the yolk. They commence foraging trips from the burrow within a week (Bassett et al. 2005) and fledge within 2 weeks at 10–16% of adult weight. Unlike other ratites they disperse after fledging and have no further contact with their parents.

2.3.1.6 General

It needs to be stated that the latter species (cassowaries and kiwis) are not likely candidates for commercial exploitation in traditional farming systems. However, subspecies of all ratite groups are threatened by dwindling populations, while some groups already became extinct (Sales 2009). The welfare of all these birds in the wild is thus under some pressure (refer Chap. 12). Major causes of a decline in numbers in wild populations are hunting as well as habitat loss to farming and other activities (Table 2.1). Farmed ratites are unlikely to become extinct, and the status of ostriches and emus are thus listed as of least concern. This is not surprising given the long history of ostrich farming in South Africa (Smit 1964; Osterhoff 1979). Commercial emu production is also well established (Scott et al. 2005). In contrast, there is a real threat to species like the kiwi and cassowary. It is therefore important to consider them as well, because the intensification of breeding may be required to stop the decline in numbers or to assist in the repopulation of suitable habitat returned to usage by wildlife (Sales 2009). Successful interventions have been reported for the kiwi (Bassett et al. 2005). Conservation management includes actions such as habitat protection by maintaining adequate reserves, control of mammalian predators and captive chick-rearing programmes. In the latter case, eggs are removed from nests in the wild, incubated at captive breeding institutions, hatched and reared to 0.8–1.2 kg, after which they are released back into the wild (Bassett et al. 2005). Barri et al. (2008) suggested that the collection and subsequent artificial incubation of "orphan eggs" (eggs not included in nests) could contribute to the conservation of the endangered lesser rhea.

It is also noted that confinement will lead to fighting because birds will have less chance to avoid competing mates. Normal courtship or mating behaviour may thus be suppressed or interrupted due to close proximity of competing mates. Excessive fighting is likely to result in injury or death in extreme situations, while the expression of normal sexual behaviour could be inhibited when paired off with an incompatible mate. Welfare of breeder birds would be compromised in both situations.

Table 2.1 Conservation status of ratites in the wild, cause for declining numbers and possible resolutions to be implemented.

Ratite group	Status	Causes for decline	Suggested action
Ostrich	Least concern	Hunting, egg gathering, overgrazing resulting in habitat loss	Captive breeding and release, genetic improvement, public awareness
Cassowary			
Southern	Vulnerable	Habitat loss and fragmentation, road kills, visitor impacts, dogs, competition and nest predation by pigs, disease, traditional demand	Habitat restoration
Northern	Vulnerable		
Dwarf	Near threatened		
Emu	Least concern	Habitat loss and fragmentation, road kills, predation of eggs and chicks by foxes, dogs and pigs	Habitat maintenance, public awareness, predator control
Rhea			
Lesser	Near threatened	Habitat loss, egg gathering, predation by felids and dogs, illegal hunting	Habitat management, control of poaching, captive breeding and release
Greater	Near threatened		
Kiwi			
Brown	Endangered	Predation	Predator control, captive breeding and release, establishment of sanctuaries, public awareness
Tokoeka	Vulnerable		
Little spotted	Vulnerable		
Great spotted	Vulnerable		

Source: Condensed from Sales (2009)

2.3.2 *Mate Choice*

Ostriches are known to have a promiscuous mating system as discussed previously, in which most individuals will have multiple sexual partners (Deeming and Bubier 1999; Davies 2002; Kimwele and Graves 2003; Bonato 2008). However, the study of Deeming (1996) indicates a definite preference in males for specific mates in some instances. If an ostrich male is paired off with an incompatible mate, eggs produced by the pair could be infertile (Malecki and Martin 2003), while the welfare of the female could be jeopardised by trauma or physical injury stemming from overt aggression by the male. Freedom to express normal sexual behaviour would also be compromised. Despite the obvious importance of mate compatibility, there is a lack of information on the topic.

Attempts to better understand this complex issue have only recently started through identification of sexual characteristics in male ostriches that prompt investment in eggs of a higher weight by females (Bonato et al. 2009a). The vividness of the white feathers and the contrast between black and white feathers in males were the most important cues for females to invest in egg weight. In contrast to popular belief, the red coloration of the shins, neck and beak was shown not to be important in this context. The authors explained that red-coloured areas are not visible when squatting down during the kantling display of males towards females. Red coloration on the shins of ostrich males also had a limited variability and almost zero heritability (Lambrechts and Cloete 2009 unpublished). The widely held belief that red shins of ostrich males indicate reproductive success may thus be false.

The colouration of ostrich males (white and black feathers, bill, neck and shins) is associated with different components of male immune capacity (Bonato et al. 2009b). These cues may signal a strong immune signal to females (that their investment in choosing a specific male as a sexual partner would be relatively safe) and to rival males (that a specific male would be a formidable opponent). These results may find application in ostrich breeding, but further development in ostriches as well as other ratites is clearly needed.

In the emu, the importance of mate choice has been recognised but not yet quantified. The practise of choosing mates by a farmer for keeping breeding birds in pairs was often counterproductive because incompatible pairing resulted in infertility of eggs, poor egg production and within-pair fighting. Given that the female is generally the larger sex, the physical advantage the female has over a male would result in a physical harm to a male, thus impacting on animal welfare. It was realised that emus need to be given an opportunity to choose partners, and nowadays most breeding emus are maintained in colonies giving them the freedom of choosing a mate and expression of normal courtship behaviour. It remains, however, unclear what male characteristics are important to a female and this is complicated by the fact that emu males and females look alike to a human eye. No male secondary sexual characters are evident. Interestingly, the female emu chooses at least one male to father her offspring while another male becomes the most suitable male to incubate and rear her chicks. The sexual cues emus use to choose their mates need to be determined to enable better management of breeding colonies.

2.4 Ratite Welfare in the Past

The South African ostrich industry is arguably the oldest ratite farming enterprise. According to Osterhoff (1979), the first ostriches were tamed in 1863. This was after proclamations were made by the then Cape Colony to protect wild ostriches from hunting for their feathers. Early breeding was mostly preoccupied with the production of the best quality feathers. The early literature therefore emphasises the evaluation of ostrich feathers and the crosses made to produce the most desirable feather quality (Smit 1964; Osterhoff 1979). The advent of incubators to incubate eggs artificially in the late 1800s led to a rapid expansion in ostrich numbers. Commercial ostrich farming for feathers soon became very popular because of a limited capital outlay and reduced labour needs. Good quality feathers could be produced by birds of up to 35–40 years (Deurden 1910; Osterhoff 1979), while extensively managed birds produced higher quality feathers than contemporaries managed intensively (Osterhoff 1979). Even though Deurden (1910) reported that feather production is optimal at 3–12 years of age, the ease of a commercial production system based on adult birds for an annual feather harvest is evident.

The welfare of the breeder birds and mature birds kept for commercial production in these early systems is not described. However, it is recognised that commercial birds were not docile. The cantankerous behaviour of such birds would have increased the risk of serious injury or death during the handling process for harvesting feathers. This led to investigations into the possibility of caponising such males in an attempt to change their behaviour (Osterhoff 1979). As this proved to be a risky operation, without marked changes in the temperament of treated birds, this practise was abandoned. It could thus be imagined that the animals were put through pain and associated fear and stress during the operation, without any subsequent gain in ease of handling. Handlers were aware that ostriches should have been handled and approached carefully because of their powerful and dangerous kicking ability (Smit 1964; Osterhoff 1979). It is not hard to imagine that the welfare of those early extensively managed ostrich flocks could have been seriously compromised, by exposing them to pain, fear and stress as well as possible serious injuries, when they were occasionally brought in for the feather harvest. A feather crop could be "forced" by drawing feather quills before they were quite ripe to stimulate the growth of new feathers (Osterhoff 1979) thus shortening the production cycle. Clearly, a practise like this would compromise welfare by causing discomfort, fear, pain and stress. At present it would therefore not be acceptable as a good agricultural practice.

Detailed information pertaining to the handling of breeder birds in those early days is also not available. From reports of Deurden (1908) and Smit (1964), it is evident that the same basic breeding structures discussed under the following heading were already in place. Given that feathers could be assessed on both sexes at an age of ~2 years, mass selection was probably all that was needed. It is thus conceivable that many breeding birds were kept in larger colonies to minimise the cost of intensification. It is interesting to note that production figures reported for four trios by Deurden (1908) were, in fact, fairly similar in terms of output per female when compared to present performance levels.

2.5 Ratite Welfare in the Present

Two basic systems dominate commercial production of ratites, namely structured breeding groups and breeding colonies. Breeding groups are typically maintained in small paddocks, while colonies may be maintained in relatively small paddocks or in larger camps with natural vegetation. According to the literature, average time budgets and reproductive behaviour are remarkably similar in the wild as well as in all these systems (Deeming and Bubier 1999).

Given their importance, these dominant production systems will be discussed next. Special mention will be made of social interactions, which can be argued to control welfare of the breeding birds, while also impacting on the commercial output from them. It is recognised that mate incompatibility in all species could result in aggressive interactions between mates. Such interactions are seen to have the potential to compromise welfare by resulting in fear and stress in subordinate birds, as well as potential injury.

2.5.1 Small Groups

2.5.1.1 Levels of Performance in Breeding Pairs (Fig. 2.2)

A breeding pair involves the mating of a single male to a single female (Smit 1964). The output in terms of eggs and chicks of pairs and trios (a single male mated to two females, discussed in the next section) is summarised in Table 2.2. Expressed relative to the period of active production (duration of lay), outputs in the literature ranged from 1.7 to 2.6 eggs per week and from 0.47 to 0.96 for chicks per week. Corresponding outputs over the entire breeding season ranged from 0.55 to 1.38 eggs per week and from 0.68 to 0.73 chicks per week. It is recognised that the pair-mating system requires a male for each female and is therefore more expensive to operate in terms of breeder maintenance (Cloete et al. 2002). As Table 2.2 shows, trios would produce more eggs on per female basis than pairs, pointing to trios as

Fig. 2.2 Ostrich pair in their breeding enclosure at the Oudtshoorn Research Farm (Photo – A. Engelbrecht)

Table 2.2 Standardised egg production from farmed ostriches in pairs and trios, as derived from the available literature. Production performance is expressed as performance per week in terms of the entire breeding season or relative to the period over which eggs were produced (duration of lay – defined as the date of the first egg subtracted from the date of the last egg produced)

Reference, breeding structure and trait	Period	Total output of structure	Days	Output per female per week
Deeming (1996)				
Pairs – egg production	Duration of lay	28.3	126	1.71
Trios – egg production	Duration of lay	67.7	123	1.90
Pairs – chick production	Duration of lay	7.0	126	0.47
Trios – chick production	Duration of lay	24.8	123	0.61
Bunter et al. (2001)				
Pairs – egg production	Duration of lay	51.1	173	2.07
Pairs – chick production	Duration of lay	23.8	173	0.96
Lambrechts et al. (2004)				
Pairs – egg production	Entire breeding season	25.9	210	0.86
Trios – egg production	Entire breeding season	52.1	210	0.87
Pairs – chick production	Entire breeding season	21.9	210	0.73
Trios – chick production	Entire breeding season	42.6	210	0.71
Cloete et al. (2005)				
Pairs – egg production	Duration of lay	44.7	167	1.87
Pairs – chick production	Duration of lay	22.1	167	0.93
Cloete et al. (2008a)				
Pairs – egg production	Entire breeding season	46.3	235	1.38
Pairs – chick production	Entire breeding season	22.9	235	0.68
Sebei and Bergaouni (2009)				
Pairs – egg production	Duration of lay	35.8	96	2.61

a better breeding unit than a pair. However, only breeders that are maintained in pairs are able to demonstrate their genetic potential for reproduction (Cloete et al. 2008b). In such a system, heritability has been estimated at 0.11–0.28 for egg production and 0.11–0.22 for chick production. Linked to very high levels of phenotypic variation (coefficients of variation exceeding 50%), these moderate heritability estimates pave the way for substantial genetic gains in chick output in a selected line compared to birds in an unselected control line (Cloete et al. 2008a) (Fig. 2.3). Selection resulted in a line of ostriches with an averaged breeding value of ~10 chicks per annum in 2006. The averaged predicted breeding values of a control line that was introduced in 1996 were relatively stable at ~2 chicks per annum since formation. Progeny of pair-bred ostriches also have pedigree information linked to production records that are needed to predict breeding values for maximising genetic progress in other important traits, as reviewed by Cloete et al. (2002, 2008b).

Pair-bred emu females are expected to produce at least 25 eggs in a season, with lower levels of production being expected from younger females (Scott et al. 2005). The range recorded in Australia indicates that a pair-bred female can lay as few as 8 and as many as 68 eggs in a season. Similarly, egg production in India averaged 30 eggs per female per annum, with ranges of 10–65 eggs per female

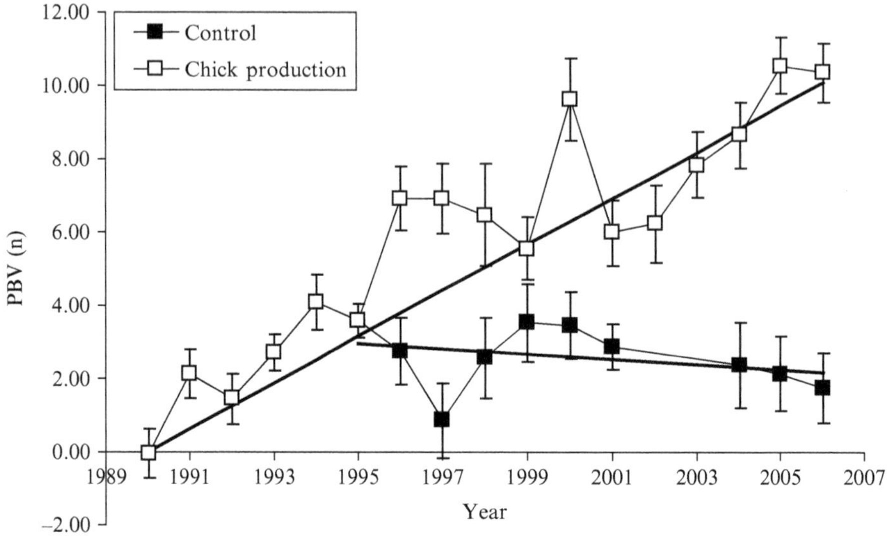

Fig. 2.3 Genetic trends, as depicted by averaged annual predicted breeding values (PBV), in an ostrich line selected for chick production and the Control line run alongside. Standard errors are denoted by *vertical lines* about the means.
Source: Cloete et al. (2008a)

(Narahari et al. 2008). These figures were derived from pairs, trios as well as breeding colonies, but unfortunately no differentiation was made between the different breeding systems. Australian breeding stock is reported to have a low turnover rate, as birds are still capable of a good reproduction after being in the system for 10 years (Scott et al. 2005). As is the case with other ratites, emus are known to easily attain ages of 20–30 years. In some cases, breeders may be evaluated in pairs before being introduced to the breeding flock (Scott et al. 2005). Birds captured from the wild cannot be sold commercially, and all commercial breeding stock has been bred in captivity (Scott et al. 2005).

It was reported that farmed greater rheas produced more eggs, also at a higher overall fertility, than semi-captive or wild birds (Navarro and Martella 2002). Captive greater rhea females produced 40 eggs per female on average compared to 24 eggs in semi-captive females. Hatching success in greater rheas is also improved in captivity (60%) compared to 45% in semi-captivity and 30% in the wild (Navarro and Martella 2002). The latter trend was reversed in lesser rheas, with a 51% hatching success in captivity compared to 60% in the wild. The improved performance of greater rheas in captivity was related to a more stable nutrient supply and the exclusion of predators (Navarro and Martella 2002). In contrast, dietary inadequacies were blamed for the suboptimal hatching performance of lesser rheas. The average egg production of 7 lesser rhea females was 32.6 eggs over a laying period of 137 days, with a range from 14 to 44 eggs (Sarasqueta 2005). A coefficient of variation for egg production of ~33% suggests marked variation between females, as is also experienced in ostriches (Cloete et al. 2002, 2008a, b).

2.5.1.2 Losses of Mature Breeding Birds in Pairs

It could be argued that breeder welfare is unlikely to be severely compromised under conditions where individuals are able to demonstrate their genetic superiority, and worthwhile genetic gains are demonstrated. However, there is no reason for complacency, as it must be conceded that the pair-mating regime does not consider the important welfare aspect of mate compatibility, as suggested previously. Mate compatibility, as mentioned by Deeming (1996), is likely to be dependent on a complex interaction of behavioural and temperamental attributes of individuals, which are poorly understood at present. In commercial farming operations, compatibility between established ostrich pairs is recognised and they are commonly retained together in the same breeding paddock over long periods of time (sometimes up to decades, given a productive life of 35–40 years) (Osterhoff 1979; Sales 1999). However, pairs used for breeding research and the demonstration of genetic gains are required to be switched regularly, to assist in the partitioning of male and female variances (Bunter 2002; Cloete et al. 2002, 2008b). Unlike with commercial operations where animals may be retained for breeding for periods of more than 20 years, birds maintained for acquiring optimal genetic gains are kept in the breeding flock for shorter periods. The reason for this deviation from traditional farming operations stems from the need for a shorter generation interval, as well as the well-defined influence of age upon hatchability, shell deaths and chick output (Bunter 2002; Cloete et al. 2006; Brand et al. 2007). Each such switch may involve an element of risk, as incompatible partners could be combined. This would conceivably lead to stress owing to overt aggression by the male, which could seriously impact upon the welfare of the female.

In view of the above arguments, data are presented for the frequency and cause of exit of male and female breeding birds from the pair-mated breeding flock at Oudtshoorn during the mating season (Table 2.3). Data were available from the 2000 to the 2009 breeding seasons. Birds that were lost during the breeding season

Table 2.3 The percentage of breeding ostriches in a pair-mating system exiting from the breeding operation on a yearly basis for the period from 2000 to 2009 in the Oudtshoorn breeding flock.

Year	Number of pairs	% Birds exiting per breeding season	
		Males	Females
2000	136	2.21	5.88
2001	136	4.41	4.41
2002	188	4.26	2.66
2003	188	1.06	1.60
2004	188	2.13	3.19
2005	188	1.60	2.13
2006	188	0.53	3.72
2007	188	2.13	1.06
2008	188	0.53	0.53
2009	188	2.66	1.06
Overall (± s.e.)	1,776	2.15 ± 0.43	2.63 ± 0.54

s.e. Standard error

Source: Brand and Phister (2010 Unpublished data)

averaged 2.2% for males and 2.6% for females. Annual losses were as low as 0.5% in both sexes (2008 in both sexes and 2006 in males), and reached maxima of 4.4% in males (2001) and 5.9% in females (2000). The relatively low level of losses is not surprising, as it is well known that mature ostriches are relatively robust and may easily grow quite old (Smit 1964; Osterhoff 1979). However, when the causes of incapacity were considered in 25 males and 37 females with data, it was clear that trauma was involved in the vast majority of cases (68% in males and 76% in females; Table 2.4). Injuries sustained by males during egg collection routines accounted for the exit of seven males (28%). In these cases, exceedingly aggressive males were either culled or permanently incapacitated in self-defence by egg collectors. The remainder of losses due to trauma stems from injuries sustained when running into fences or other objects in both males and females, except for one female that died because of an internal egg breakage. Losses of minor importance included deaths due to excessive heat that amounted to 16% of cases in males and 8.1% of cases in females. Corresponding percentages of unspecified losses were 8% and 14%. One female died of respiratory disease, one male was killed when a dietary fragment pierced its intestines and one male was killed by snakebite. It is clear that trauma (also induced by the handlers in the case of male birds) was the single most important cause of losses of mature breeding birds in the mating season. Smit (1964) accordingly stated that mature ostriches seldom die of natural causes, but that they were often incapacitated because of injury sustained because of their wild nature.

Against the background of 76% of females exiting from the breeding flock because of trauma, there is an obvious problem with adaptability of males to routine egg collection. Stress so induced may also affect females and compromise egg quantity and quality, as well as fertilisation rate, but experimental evidence is still lacking. There is an obvious disadvantage associated with an incompatible pair in a small paddock. The female is likely to have limited possibilities to flee from an aggressive mate and is more likely to be injured or killed relative to contemporaries in groups and on larger areas, where escape should arguably be easier. The stress of being paired off in a confined space with an incompatible mate could also result in self-inflicted injury by, for instance, running into the fence or other obstacles when threatened. At the very least mate compatibility would compromise the welfare of animals by impacting on the freedom to express normal sexual behaviour.

Table 2.4 The overall causes of exit of breeding ostriches in a pair-mating system for the period from 2000 to 2009 in the Oudtshoorn breeding flock, excepting 2002 and 2007 when data were not available.

Number of birds and cause of exit	Gender	
	Male	Female
Total number of birds	25	37
Trauma	17 (68.0%)	28 (75.7%)
Disease	0 (0.0%)	1 (2.7%)
Climate	4 (16.0%)	3 (8.1%)
Dietary	1 (4.0%)	0 (0.0)
Snakebite	1 (4.0%)	0 (0.0)
Unknown	2 (8.0%)	5 (13.5%)

Source: Brand and Phister (2010 Unpublished data)

If the contribution of trauma is subtracted from annual losses, it is clear that natural attrition is almost negligible and in some cases purely coincidental. If the welfare of farmed ostrich breeders in pairs is to be addressed, conditions leading to trauma needs to be identified, quantified and eliminated. Of these, mate compatibility seems to be first and foremost, while husbandry practices and male aggression towards egg collectors should also be addressed.

We are not aware of comparable figures as pertaining to the levels and causes of mortality of adult breeding birds for other farmed ratites (rheas and emus). The commencement of production in captivity in both these groups of birds has been fairly recent. Farming with the greater rhea got underway in the early 1990s (Navarro and Martella 2002), while farming with the lesser rhea commenced in 1994–1995 (Navarro et al. 2003). It could therefore be argued that a similar scenario as for farmed ostriches may be likely. Emus were only recognised as farm animals in Western Australia in 1987, while the other Australian states followed suit by 1994 (Scott et al. 2005). However, farmers quickly realised that artificial pairing could lead to bird fighting, injury and poor egg production if incompatible birds were put into the same enclosure. Nowadays, the practise of artificial pairing is abandoned and replaced by natural pairing in larger colonies. Once a compatible pair is identified it is then captured and moved to a separate enclosure should the individual egg record be required. The relative smaller size of emus and rheas may result in them being less prone to death and serious injury because of collisions with solid objects.

2.5.1.3 Mature Breeding Birds as Foster Parents (Fig. 2.4)

Ratite chicks raised in captivity often suffer from high levels of mortality, with mortality rates sometimes approaching or exceeding 80% in the greater rhea (Labaque et al. 1998) and in the ostrich (Cloete et al. 2001). These high levels of attrition hold obvious welfare implications for the farmed ratite industries. The adoption of chicks by males in the rhea and by breeding pairs in the ostrich may play a role to alleviate the problem of excessive chick deaths. Greater rhea chicks adopted by males had a similar survival to chicks reared intensively (0.47 vs. 0.43) but maintained a better growth rate up to ~90 days of age (Barri et al. 2005). In another study, the survival of adopted chicks was reported to be similar than that of chicks hatched by foster males (Labaque et al. 1998). The survival of ostrich chicks reared with breeding pairs was almost double that of chicks reared intensively (0.82 vs. 0.50) (Janse van Vuuren 2008). In this case, welfare is promoted by reduced levels of suffering and distress in chicks as well as by allowing foster parents to express normal behaviour involving parental care. Survival of emu chicks is generally good, so foster parenting has hardly been practised. Emu chicks could only be looked after by the male (as for rheas), which is the sole incubator and also responsible for parental care for more than a year under natural conditions.

Fig. 2.4 Ostrich pair fostering chicks at the Oudtshoorn Research Farm (Photos – (a) I.A. Malecki and (b) A. Engelbrecht)

2.5.1.4 Levels of Performance in Other Breeding Structures Involving Small Groups

It is evident that egg production per female is not compromised in trios (Table 2.2). Corresponding figures for pairs are provided in brackets to facilitate comparisons in the following text. Expressed relative to the period of active production (duration of lay), outputs of trios in the literature were 1.9 eggs per week (1.7–2.6 eggs per week in pairs) and 0.71 chicks per week (0.47–0.96 for chicks per week in pairs). Corresponding outputs over the entire breeding season ranged from 0.35 to 0.87 eggs per week (0.55–1.38 eggs per week in pairs) and were 0.71 chicks per week (0.68–0.73 chicks per week in pairs). Naturally the production per enclosure should be greater in females forming part of a trio, as more females are able to contribute. Results of quads (one male with three females) are scant in the literature. Relatively favourable performance figures presented by Lambrechts et al. (2004) should be treated with caution, as it is based on comparatively few data. It is of interest that the quad structure was preferred for lesser rheas in the study of Sarasqueta (2005).

The welfare situation in trios and quads is likely to be improved relative to pairs, as the potential for aggressive interactions owing to mate incompatibility is reduced in proportion to the number of females. Stress resulting from aggression by an incompatible mate, potentially leading to serious injury or death, would be accordingly reduced. Normal sexual behaviour will also be promoted. However, the usage of a communal nest precludes the conclusive determination of pedigree information that is needed for genetic evaluation. The introduction of an affordable DNA-pedigreeing service for the ratite industries may change this situation. Based on the high repeatability of egg weight in ostriches (Bunter and Cloete 2004), this limitation can be overcome by prior knowledge of egg weight of individual females constituting the trio (Deeming 1996; Essa and Cloete 2004). Lesser rhea eggs can similarly be allocated to different females on the basis of shape, size and shell structure (Sarasqueta 2005). Emus, on the other hand, are not likely to be successful when made into trios. Females may compete strongly over a male, that results in between-female fighting in which the male may also become involved. Such interactions are usually harmful to at least one bird and a free expression of reproductive behaviour is difficult or impossible and breeding unsuccessful unless one female is removed from the paddock. A limited report from India (Narahari et al. 2008) indicates emu trios may be successful if the trio is reared together from the chick age (Narahari personal communication), but no substantial evidence has been gathered to make any valid recommendation for trio settings on emu farms.

2.5.1.5 Genetics of Temperament

In all these systems, male aggression towards egg collectors compromises breeder welfare in ostriches, while also posing a definite threat to human safety (Table 2.4). Lambrechts and Cloete (2009 unpublished) devised a monthly, subjective scoring system for male aggression towards egg collectors. These scores were averaged to obtain unique records for males within production years. When analysed these scores were moderately heritable at 0.25 (SE = 0.06) when assessed as year averages. This result demonstrates that at least one aspect of ostrich temperament was heritable and would respond to selection if needed. However, the impact of selection for a more docile temperament on the fertility of eggs produced by companion females should still be assessed. Temperament and aggression to egg collectors is possibly not as important in the farmed ratite species of smaller stature (rheas and emus) where aggression against egg collectors is less likely.

2.5.2 Colony Breeding

It needs to be recognised that approximately 80% of ostrich breeders in South Africa and elsewhere are maintained in colonies, often at a sex ratio of approximately six males for every ten females (Lambrechts et al. 2004). Because these colonies may differ vastly in size, it is necessary to differentiate between small colonies of <20 breeding females and larger colonies. These options are discussed below.

2.5.2.1 Small Colonies (Fig. 2.5)

The output in terms of eggs and chicks of small colonies is summarised in Table 2.5. Ranges of outputs that were found for pairs are also provided in brackets in the text below, to assist in comparisons. Expressed relative to the period of active production (duration of lay), outputs from small colonies in the literature ranged from 1.07 to 1.44 eggs per female per week (1.7–2.6 eggs per week in pairs) and from 0.15 to 0.48 chicks per female per week (0.47–0.96 for chicks per week in pairs). Corresponding outputs over the entire breeding season ranged from 1.01 to 1.41 eggs per week (0.55–1.38 eggs per week in pairs) and from 0.82 to 1.03 chicks per week (0.68–0.73 chicks per week in pairs).

From these results, it is clear that ostriches can successfully be housed in small groups in different enclosures, while meeting expectations for commercial production in terms of egg and chick output. Such groupings should be preferred from a welfare perspective, as it is the closest approximation of ostrich breeding groups under natural conditions (Kimwele and Graves 2003). Small colonies require a smaller capital outlay per breeding bird than pairs, trios and quads, as a 9-bird colony

Fig. 2.5 A small colony of ostriches at the Oudtshoorn Research Farm (Photo – A. Engelbrecht)

Table 2.5 Standardised egg and chick production from farmed ostriches in small colonies, as derived from the available literature. Production performance is expressed per week for the entire breeding season or relative to the period over which eggs were produced (duration of lay – defined as the date of the first egg subtracted from the date of the last egg produced)

Reference, trait and year	Number of females	Period	Output	Days	Output per week
Deeming (1996)					
Egg production					
1995	4	Duration of lay	84	126	1.17
1995	8	Duration of lay	287	175	1.44
1995	15	Duration of lay	352	154	1.07
Chick production					
1995	4	Duration of lay	11	126	0.15
1995	8	Duration of lay	96	175	0.48
1995	15	Duration of lay	84	154	0.26
Lambrechts et al. (2004)					
Egg production					
2000	9	Entire breeding season	233.7	150	1.21
2000	6	Entire breeding season	129.8	150	1.01
2001	6	Entire breeding season	255.6	210	1.41
2001	6	Entire breeding season	233.4	210	1.30
Chick production					
2000	9	Entire breeding season	198.3	150	1.03
2000	6	Entire breeding season	108.0	150	0.84
2001	6	Entire breeding season	182.6	210	1.01
2001	6	Entire breeding season	147.8	210	0.82

can be accommodated on the same area as needed for a pair or trio (Lambrechts et al. 2004). The mating group also allow for the expression of normal sexual behaviour between members of the colony as well as a reduced potential for problems associated with mate incompatibility. However, this structure precludes the routine recording of pedigree information, unless an affordable DNA-pedigreeing service becomes routinely available. In this case, differentiation between the eggs produced by different females on egg shape, size or shell structure is unlikely to be successful. The aberrant sexual behaviour reported by Lambrechts et al. (2004) for larger colonies noted in the following section do not seem to be present in smaller colonies.

2.5.2.2 Large Colonies (Fig. 2.6)

Egg and chick output per female from large colonies over the entire breeding season on a commercial farm is presented in Table 2.6. Egg output ranged from 0.79 to 1.23 eggs per breeding female per week, which compares favourably with weekly egg outputs of 0.55–1.38 eggs per week in pairs (Table 2.2). The corresponding range for chick production was between 0.28 and 0.85 chicks per female per week compared to a range of 0.68–0.73 chicks per female per week in pairs. Chick

Fig. 2.6 A colony of breeder emus in Western Australia (Photo – P. Rybnik-Trzaskowska)

Table 2.6 Standardised egg and chick production from farmed ostriches in large colonies, as derived from Lambrechts et al. (2004). Production performance is expressed as performance per week in terms of the entire breeding season

Trait and year	Number of females	Period	Output	Days	Output per week
Egg production					
2000	80	Entire breeding season	1,748	150	1.02
2000	100	Entire breeding season	2,181	150	1.02
2000	140	Entire breeding season	2,859	150	0.95
Chick production					
2000	80	Entire breeding season	515	150	0.30
2000	100	Entire breeding season	707	150	0.33
2000	140	Entire breeding season	844	150	0.28
Egg production					
2001	76	Entire breeding season	2,793	210	1.23
2001	87	Entire breeding season	2,849	210	1.09
2001	94	Entire breeding season	2,230	210	0.79
Chick production					
2001	76	Entire breeding season	1,938	210	0.85
2001	87	Entire breeding season	1,917	210	0.73
2001	94	Entire breeding season	1,510	210	0.54

production may have been compromised to an extent during 2000, but these birds were affected by a mid-season reallocation, resulting in a shortened breeding season of 150 days (Lambrechts et al. 2004) (Table 2.6). The clustering according to year is evident when the chick output is plotted against the number of breeding females present in the breeding colonies in Fig. 2.7. The chick output seemed to decline with an increase in density (from an area of 87.7 m^2 per bird to 70.9 m^2 per bird) in 2001. With our present knowledge, it is not clear whether this trend is coincidental, and further research is indicated.

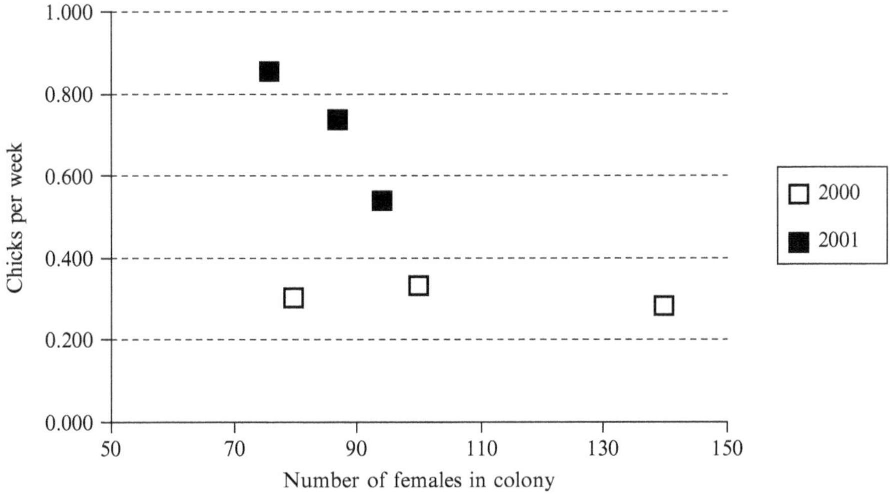

Fig. 2.7 Chick output per female per weeks in large colonies according to the number of females in the colony for 2000 and 2001. Males and females were joined at a 1:2 ratio in 1 ha paddocks. *Source*: Adapted from Lambrechts et al. (2004)

The usage of large colonies on a limited area has been reported to lead to a breakdown in the social structure of the colony (Lambrechts et al. 2004). However, some colonies of more than 100 birds each appeared to breed successfully (Ledger and Malecki 2008; I.A. Malecki unpublished). Under such conditions, there may be more than 10% of eggs laid outside nests, which are usually not fertilised (I.A. Malecki unpublished). It was suggested that the high stocking densities that were used prevented males establishing territories. The resultant fighting bouts could have disturbed normal breeding activities in other colony members. Excessive territorial aggression among competing males is likely to compromise welfare by causing serious injury or even death of some colony members. Other aberrant behaviours that were mentioned included an increase in homosexual behaviour as well as feather pecking (Lambrechts et al. 2004). The former behaviour precludes normal sexual activity to an extent, while the latter is recognised as stereotyped behaviour observed in birds subjected to chronic stress (Samson 1996).

2.5.3 Intensification Versus Extensive "Ranching"

In farmed ratites, animals can be kept under contrasting conditions. Semi-extensive ranching with rheas is an option to consider (Sarasqueta 2005). Wild and captive rheas often occur alongside each other, while greater rheas do not seem to show a strong aversion towards manmade structures and human activities (Herrera et al. 2004) in an agricultural ecosystem. Yet suitable habitat for greater rheas in the wild is under pressure, as modified rangeland is seen to earn greater incomes for landowners (Giordano et al. 2009). Welfare of ratite breeders in terms of the freedom to express their natural behaviour seems to be best served under more extensive conditions irrespective of their group size, as was found with the early South African ostrich industry (Osterhoff 1979). The available wild rhea gene pool is seen by Sarasqueta (2005) as being important to source new attributes to add to captive stock. The contrary seems to be the case in ostriches and emus. Emus to be sold as breeding material should have been bred in captivity (Scott et al. 2005). The infusion of wild stock into the domestic South African ostrich population also appears to be highly unlikely because of an undesirable temperament and low levels of production. As an example, it was shown that the more recently domesticated Zimbabwean Blue birds had a markedly reduced chick survival compared to the domestic South African Black strain (Engelbrecht et al. 2008). Meat produced by Zimbabwean Blue birds accordingly was higher in pH than that of South African Black birds (Hoffman et al. 2008). The latter result was later confirmed in a larger study, while a similar tendency was observed for the Kenyan Redneck strain (Davids et al. 2010). This result was attributed to the more recently domesticated strains being more skittish and more likely to be stressed. An increase in skittishness and unpredictability is likely to be associated with a greater probability of death and injury during routine husbandry operations. An increase in muscle pH was also related to stress inflicted by an increased lairage period in slaughter ostriches (Van Schalkwyk et al. 2005).

Breeding ostriches in South Africa are sometimes maintained in large, extensive paddocks (Lambrechts et al. 2004). It is reasonable to assume that this system would be the best for the welfare of such breeders, as it closely approximates conditions in nature. The risk of adverse social interaction between rival males and incompatible mates (as discussed previously) is likely to be minimised under such conditions, while freedom to express normal sexual behaviour is likely to be maximised. This argument is complicated by the fact that the Klein Karoo, where the bulk of South Africa's commercial ostriches are found, is part of an important international biodiversity hotspot, namely the succulent Karoo. The succulent Karoo is defined as a winter rain desert along the west coast of South Africa and Namibia, which extends into the southern Karoo (thus incorporating the Klein Karoo region – Cowling 2010). The botanical diversity of this region is unsurpassed globally by other arid regions. When breeding birds are maintained in large paddocks, it has been demonstrated that the vegetation is severely compromised near to watering and feeding points (Fig. 2.8). Grazing with ostriches has, in fact, been shown to be more detrimental to the condition of natural pasture than grazing with either sheep or cattle in the Klein Karoo area of South Africa (Cupido 2005) (Fig. 2.9). This has happened despite ostriches receiving adequate production diets when utilising large paddocks with natural vegetation. Because of this impact on the plant cover, strict legislation in the form of the Act on Conservation of Agricultural Resources (Act 43 of 1983; South African Ostrich Business Chamber 2009) has been passed to protect the natural resource. Under this act, the growing out of slaughter ostriches on the natural pastures (veld) is not permissible (South African Ostrich Business Chamber 2009). The usage of veld to keep breeding ostriches

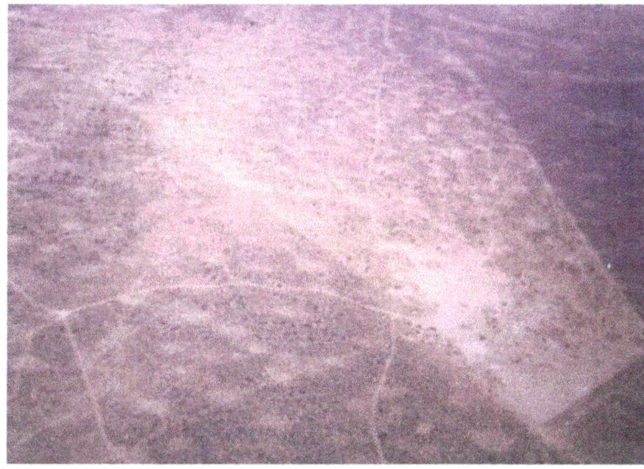

Fig. 2.8 An aerial view of ostrich breeding paddocks on natural pasture of the Succulent Karoo ecotype. The local soil degradation in the areas surrounding the feeding and watering points is clearly visible (Photo – S. Botha)

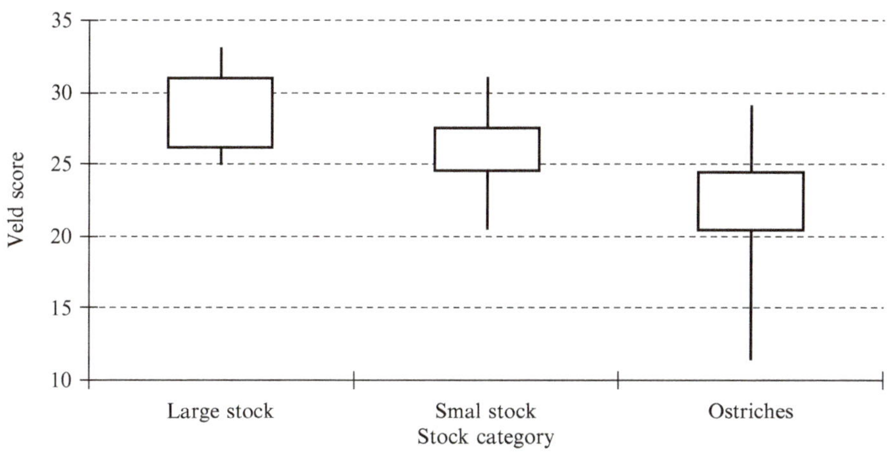

Fig. 2.9 The veld condition score of natural Karoo pasture utilised by different livestock classes. The *box* represents 95% confidence intervals and the *high–low lines* the range of values encompassed by the data.
Source: Adapted from Cupido (2005)

is allowed, on the provision that they should receive a complete breeding diet, and precautions are made to guard against veld degradation (South African Ostrich Business Chamber 2009). Against this background, there is a movement to intensify ostrich operations to ensure that the biodiversity of the area is not compromised. It is important to consider this movement in relation to the welfare of breeder birds as well as the sustainability of the natural resource.

Concentration of ostrich breeders (Lambrechts et al. 2004) on a limited area quickly leads to an eradication of the available vegetation and to the soil becoming denuded. This may result in wind erosion as well as the aberrant behaviour reported under the previous heading. This is clearly not in the best interest of breeder welfare, as the conditions may lead to conditions such as chronic respiratory disease (Huchzermeyer 1999). The intensive maintenance and recording of performance under pair-breeding conditions may be less detrimental for the natural resource. However, as stated previously, it may compromise breeder welfare by mate incompatibility. Capital expenditure may also be a major deterrent. Cloete et al. (2002, 2008b) recommended this outlay only for breeding operations that are serious in achieving genetic gains in their stock and willing to supply the additional capital that is needed. The best compromise for commercial producers would be small colonies, as in Table 2.5. Such a system would be best synchronised with the natural ostrich reproductive behaviour, thus benefiting breeder welfare in terms of mate compatibility (with the welfare benefits listed previously) as well as freedom to express normal sexual behaviour. It will also be less expensive to maintain. As feed and watering points should always be near, it may have a smaller environmental impact in the long run than larger colonies on a confined space. However, it needs to be stated that animal needs in terms of group sizes, stocking density and male:female ratio have not been explicitly studied across predefined ranges.

Currently, in Australia at least, emus are mainly maintained in colonies of between 50 and 200 breeding birds in large paddocks with variable edible vegetation for emus, supplemented by concentrated rations. Emu colony size is such that birds appear to have sufficient space to establish nest sites and territories and fighting is not of concern to the farmers, as they are in general of a natural occurrence as mate and site competitive interactions. Provided that sufficient space is available, subordinate animals are usually able to retire to safety before being seriously injured.

2.6 Future Perspectives

Mitchell (1999) reported a lack of scientific studies on the welfare of ostriches in 1999. Very few studies accrued since, while the literature on the other ratite species is even more scant. However, two farmed ratite species have Codes of Conduct to regulate their gross welfare needs (Primary Industries Ministerial Council 2006; South African Ostrich Business Chamber 2009). The compilation of Codes of Conduct is only the beginning of a number of steps that need to be taken.

Ethical concerns dictate that the welfare of farmed animals under routine husbandry operations should always be considered. Temperament is thus important in all farmed livestock species, and a lack of undue fear for humans should accompany all routine husbandry operations (Boissy et al. 2002). Jensen et al. (2008) reviewed the genetics and genomics of animal behaviour and welfare in livestock. Several classical examples were provided of how desired outcomes in terms of tameness of foxes, fear in mink and feather pecking in chickens were overcome by directed selection. Yet there are no such good-news stories in the available literature on farmed ratites, and we are not familiar with studies in progress where the temperament of farmed ratites is studied explicitly. The reported genetic variation in aggressiveness of male ostriches (Lambrechts and Cloete 2009 unpublished) is of interest and deserves further study. There are a number of other aspects of welfare that should be investigated for genetic variation. These include, for example, adaptation to routine husbandry procedures, lower levels of fear and skittishness when startled and social behaviour. Genetic analyses in future should investigate whether the inclusion of social interactions in the genetic model, as propagated by Bijma (2009), will hold any advantage in analysing the data of the naturally gregarious farmed ratites. It is foreseen that this may be the case under severe competition for resources.

The factors involved in mate choice also deserve further study, to benefit breeder welfare by ensuring that compatible individuals are combined. In other livestock species, temperament and behaviour are generally unrelated to quantitative production traits (Burrow 2001). However, this needs to be established in ratites.

Further research should consider linking phenotypes to a high density genetic map for the species to attempt unravelling the genomics of behaviour and welfare in ratite species. For this to be efficient, genomic information needs to be related to phenotypic data on measures of behaviour and/or temperament that still need to be defined. The importance of linking genomic information to phenotypes has recently been stressed by Goddard et al. (2010). This has to be seen as a major if somewhat idealistic objective, but with limited immediate application, as high-density genetic maps are not available for any of the farmed ratite species. To our knowledge, the most comprehensive molecular genetic information for ratites to date is a set of microsatellite markers reported for ostriches (Huang et al. 2006, 2008). Yet there are known quantitative trait loci for temperament and locomotion in cattle, as well as for fearfulness, adaptability and feather pecking in chickens (Jensen et al. 2008). The constant improvement of genomic knowledge and techniques is reassuring in this respect. Studies that may seem farfetched at present may become feasible with the advent of improved technology. Yet, to enable these advances, the ratite industries need to be willing to invest in research and development in these fields. It may be that the farmed ratite industries are not stable enough at present to invest in this way.

In the process of compiling this review, we came across discrepancies within the ratite literature. For instance, Barri et al. (2005) argued that the reproduction of prospective greater rhea males may be enhanced by adoption and a lack of human imprinting. Conversely, human imprinting is considered indispensable for the development of a viable artificial insemination protocol for other ratite species (Malecki et al. 2008). The role of imprinting on humans on the reproduction and survival of ratites should thus be the topic of further studies. The underlying mechanisms controlling ovulation in commercially farmed ratites (emus, ostriches and rheas) needs to be understood if we are to achieve the most efficient male: female ratio. However, these species are likely to pose different challenges because of differences in their mating systems. While female emus rarely accept each other in close proximity, ostrich and rhea females tolerate being in a group and contribute eggs to the same nest.

It was interesting to note that the same basic procedures for the handling of ostriches described by Smit (1964) are still firmly in place almost 50 years later.

Breeding birds will not produce optimally if their welfare is seriously compromised. Still there are many aspects of ethical ratite production that are well below what should be attainable. Trauma as the major cause for mature ostriches exiting the breeding flock (Table 2.4) prematurely is clearly not advisable from an animal ethics and welfare perspective. There is thus an urgent need for conducting research on the temperament of these birds under farming conditions. A continued lack of adaptation of ostriches and other ratites to their production environment as well as a lack of consideration of their specific behavioural patterns would be unacceptable.

2.7 Conclusions

This chapter reviews aspects of welfare of ratite breeders under farming conditions. In the process, we considered the reproductive behaviour and strategies of ratites as well as factors important in mate choice. Past tendencies were considered, while an account of the present situation was given. The pros and cons of different mating groups from pairs to large colonies were considered, in relation to animal needs, as well as natural biodiversity. Lastly, we considered areas of behaviour and welfare that have not yet been addressed by ratite breeders. These include interactions with other birds and/or humans, as well as the freedom to express normal behaviour, and predominantly sexual behaviour. Some of the noted gaps in scientific knowledge may be addressed soon, within the constraints of the presently available manpower and monetary allocations. Other interventions were considered as more ambitious and would need a significant injection of capital to become reality. If the recommendations could all be followed, it would pave the way for challenging and exciting studies on genetics, behaviour and social interactions of all the major ratite groupings, an aspect that has been seriously neglected so far.

References

Barri FR, Navarro JL, Maciera NO, Martella MB (2005) Rearing greater rhea (*Rhea Americana*) chicks: is adoption more effective than the artificial intensive system? Br Poult Sci 46:22–25

Barri FR, Martella MB, Navarro JL (2008) Characteristics, abundance and fertility of orphan eggs of the Lesser Rhea (*Pterocnemia-Rhea-pennata pennata*): implications for conservation. J Ornithol 149:285–288

Bassett SM, McLennan JA, Blackwell GL (2005) The biology and conservation status of the brown kiwi. In: Carbajo E (ed) Proceedings of the 3rd International Ratite Science Symposium and XII World Ostrich Congress, Madrid, Spain, 14–16 Oct 2005, pp 165–169

Bijma P (2009) Maintaining fitness by within breed selection. In: Van der Werf J, Graser H-U, Frankham R, Gondro C (eds) Adaptation and fitness in animal populations. Springer Science and Business Media, Berlin, pp 103–124

Blache D, Barrett CD, Martin GB (2000) Social mating system and sexual behaviour in captive emus, *Dromaius novaehollandiae*. Emu 100:161–168

Malecki IA, Martin GB (2003) Sperm supply and egg fertilization in the ostrich (*Struthio camelus*). Reprod Dom Anim 38:429–435
Malecki IA, Rybnik PK, Martin GB (2008) Artificial insemination technology for ratites: a review. Aust J Exp Agric 48:1284–1292
Mitchell MA (1999) Welfare. In: Deeming DC (ed) The ostrich, biology, production and health. CAB International, Wallingford, pp 215–230
Moore LA (2007) Population ecology of the southern cassowary *Casuarius* casuarius johnsonii, Mission Beach north Queensland. J Ornithol 148:357–366
Narahari D, Ramamurthy N, Kumararaj R, Gnanaraj PT (2008) Present status of emu farming in India. Aust J Exp Agric 48, xix. In: 4th International Ratite Science Symposium, mini paper and poster booklet, 29 June–4 July 2008, Brisbane, Australia. http://www.publish.csiro.au/?act=view_file&file_id=EAv48n10posters.pdf. Accessed 1 Oct 2010
Navarro JL, Martella MB (2002) Reproductivity and raising of greater rhea (*Rhea Americana*) and lesser rhea (*Pterocnrmia pennata*): a review. Archiv für Geflügelkunde 66:124–132
Navarro JL, Barri FR, Maestri DM, Labuckas DO, Martella MB (2003) Physical characteristics and chemical composition of Lesser Rhea (*Pterocnemia pennata*) eggs from farmed populations. Br Poult Sci 44:586–590
Osterhoff DR (1979) Ostrich farming in South Africa. World Rev Anim Prod 15:19–30
Primary Industries Ministerial Council (2006) Husbandry of captive-bred emus, 2nd edn. PISC report no 19. Model code of practice for the welfare of animals. Available from: http://www.publish.csiro.au/Books/download.cfm?ID=5390. Accessed 1 Oct 2010
Reid B, Williams GR (1975) The kiwi. In: Kuschel G (ed) Biography and ecology in New Zealand. Junk, The Hague, The Netherlands, pp 301–330
Ridley M (1978) Paternal care. Anim Behav 26:904–932
Sales J (1999) Slaughter and products. In: Deeming DC (ed) The ostrich, biology, production and health. CAB International, Wallingford, pp 231–274
Sales J (2006) The rhea, a ratite native to South America. Avian Poult Biol Rev 17:105–124
Sales J (2007) The emu (*Dromaius novaehollandiae*): a review of its biology and commercial products. Avian Poult Biol Rev 18:1–20
Sales J (2009) Current conservation status of ratites. J Threat Taxa 1:9–16
Samson J (1996) Behavioral problems of farmed ostriches in Canada. Can Vet J 37:412–414
Sarasqueta DV (2005) Aspects of rearing, reproduction and hybridization of Darwin's rhea or choique (*Rhea pennata* syn. *Pterocnemia pennata*, spp *pennata*). In: Carbajo E (ed) Proceedings of the 3rd International Ratite Science Symposium and XII World Ostrich Congress, Madrid, Spain, 14–16 Oct 2005, pp 35–44
Scott P, Turner A, Bibby S, Chamings A (2005) Structure and dynamics of Australia's commercial poultry and ratite industries. Report prepared for The Department of Agriculture, Fisheries and Forestry by Scolexia Animal and Avian Health Consultancy, 16 Learmonth Street Moonee Ponds Victoria, Australia 3039
Sebei SK, Bergaouni R (2009) Ostriches' reproduction behaviour and mastery of natural incubation under farming conditions. Trop Anim Health Prod 41:353–361
Smit DJvZ (1964) Ostrich farming in the Little Karoo. Pamphlet no. 358. Department of Agricultural Technical Services, Pretoria, Republic of South Africa
Smith JM (1977) Paternal investment: a prospective analysis. Anim Behav 25:1–9
Soley J, Groenewald HB (1999) Reproduction. In: Deeming DC (ed) The ostrich, biology, production and health. CAB International, Wallingford, pp 129–158
South African Ostrich Business Chamber (2009) Code of conduct for the commercial production of ostriches, 4th edn. Updated version October 2009. http://www.ostrichsa.co.za/downloads/code_of_conduct_oct_09.pdf. Accessed 1 Oct 2010
Taylor EL, Blache D, Groth D, Wetherall JD, Martin GB (2000) Genetic evidence for mixed parentage in nests of emu (*Dromaius novaehollandiae*). Behav Ecol Sociobiol 47:359–364
Van Schalkwyk SJ, Hoffman LC, Cloete SWP, Mellett FD (2005) The effect of feed withdrawal during lairage on meat quality characteristics in ostriches. Meat Sci 69:647–651

Chapter 3
Natural Mating and Artificial Insemination

I.A. Malecki and P.K. Rybnik-Trzaskowska

Abstract Understanding normal reproductive behaviour of ratites encourages more efficient farming and improves the sustainability of captive and wild populations. Providing freedom of expression of normal sexual behaviour ensures that birds naturally select their partners, which should increase the success of breeding under natural mating. However, from the farmer's perspective sexual selection may not always be desired because it may not result in the selection for economically important traits. Artificial selection of mates may result in incompatible pairings, welfare problems and poor fertility. Selection for economic traits is needed to improve efficiency of farming ratites and artificial insemination technology could facilitate the fastest rate of improvement. Welfare friendly methods for semen collection and artificial insemination have been developed for this purpose whereby normal sexual responses evoked by sexual stimuli lead to sexual crouch, voluntary ejaculation and non-intrusive artificial insemination. The lack of fear or aggression in some birds and imprinting to humans in response to stimuli have resulted in the development of positive human–ratite relationship during artificial insemination and semen collection. The expression of friendly and sexual behaviour of birds towards humans may have welfare implications, but it could also lead to tamer birds, more efficient production of ratites and a good animal–human relationship.

Keywords Imprinting · Semen collection · Sexual behaviour · Training · Welfare

3.1 Introduction

Efficient production of ratites is constrained by the breeding biology of the species, variation in adaptability to farming environment and variation in production traits.

I.A. Malecki (✉) and P.K. Rybnik-Trzaskowska
School of Animal Biology MO92, The University of Western Australia, Faculty of Natural and Agricultural Sciences, Crawley, WA 6009, Australia
and
UWA Institute of Agriculture, The University of Western Australia, 35 Stirling Highway, Crawley, WA 6009, Australia
e-mail: irek.malecki@uwa.edu.au; prybnik@gmail.com

P.C. Glatz et al. (eds.), *The Welfare of Farmed Ratites*, Animal Welfare 11,
DOI 10.1007/978-3-642-19297-5_3,

One strategy to improve the efficiency of ratite farming is a structured breeding and selection program that uses artificial insemination (AI) technology. Its development, however, has for some time been hampered by the lack of reliable semen collection and artificial insemination methods. Attempts to develop AI for ostrich farming were made more than 30 years ago, when semen collection and artificial insemination methods were proposed (Von Rautenfeld 1977). The methods were based on the physical restraint of birds. Little consideration was given to the bird's welfare during their capture and restraint, semen collection and during the artificial insemination procedure. It was not realised (until very recently) that the freedom to express normal sexual behaviour and freedom from fear and stress were not being considered. These freedoms are now regarded as very important for the artificial insemination technology to be successful and acceptable as a procedure. Capture and physical restraint of ostriches, emus or rheas result in considerable stress. It creates a risk of injury to the birds that may lead to poor reproductive output through inhibition of ovulation, poor quality of ejaculates and poor fertility in addition to the stress and risk of injury to the birds and their handlers. During a relatively short history of ratite farming little scientifically-based selection has been done to adapt ratites to their farming environment. While some individuals show a wild nature, others are calm, showing little fear of humans or expressing friendly behaviours to humans. Identification of suitable individuals and allowing them to express sexual behaviour towards their conspecifics or humans have allowed the development of methods for collecting semen from emus (Malecki et al. 1997a) and ostriches (Rybnik et al. 2007), and for artificial insemination of emus (Malecki and Martin 2004) and ostriches (Malecki et al. 2008). In this chapter, we discuss how ratites express their sexual behaviour and show that the same behaviour patterns that are observed during natural mating are also observed during semen collection and artificial insemination. Our experience is that understanding ratites' behaviour helps to improve human–bird relationship, provides better welfare for the birds, good working conditions and an improved farming environment. On the other hand, if there is a lack of understanding of their sexual behaviour or their need to express it freely, it may have serious implications for their welfare and farm output. Recently developed novel approaches in handling ratites for the purpose of semen collection and artificial insemination do not appear to compromise their welfare and may provide a solution for development of the new ratite industry and effective conservation strategies. Increasing the essential knowledge base of ratites reproductive behaviour and creating an optimal farming environment allows free expression of natural sexual behaviour while welfare-improved methods of artificial insemination can improve bird welfare.

3.2 Natural Mating

The free expression of the sexual behaviour in captive-bred ratites is the key to successful mating and sustained fertility. Currently, ratites are bred by natural mating only and egg fertility is satisfactory (Malecki and Martin 2002a, 2003).

At times there are insufficient conditions provided for birds to allow full freedom of expression of sexual behaviour. Our understanding of the behaviour of ratites and the conditions they require to fully express their sexual behaviour are still poor. In addition, interference may deprive ratites of free expression and mate choice. In captivity, there may be more inhibiting factors than the birds would normally encounter in the wild. These include human presence, enclosure size and type and/or husbandry practices (Bubier et al. 1998). Activities on farm may disrupt mating or may facilitate imprinting that may also be undesirable. On the other hand, human appearance and activities may stimulate the expression of sexual behaviour (Bubier et al. 1998; Rozenboim et al. 2003; Malecki et al. 2008). We are yet to determine what the optimal environment for the ratites is, so up to this point, we assume that under the conditions we provide, the birds do allow expression of most of the behaviours.

3.2.1 Expression of Normal Sexual Behaviour

The sexual behaviour has been described in free-range and captive-bred emus (Malecki 1993; Coddington and Cockburn 1995; Malecki et al. 1997a; Blache et al. 2000) and ostriches (Sauer and Sauer 1966, Bertram 1979; Bubier et al. 1998, Rybnik et al. 2007). It has been shown that successful mating constitutes a co-ordinated series of behavioural acts or a series of responses to specific stimuli that can be allocated into one of four phases: courtship, mounting, intromission and ejaculation and post-coital display. In each phase, the stimuli eliciting sexual responses in a female and in a male ultimately lead to the next phase until they culminate in successful semen deposition, semen retention and egg fertilization.

3.2.1.1 Stimuli Evoking Sexual Response in a Female

Courtship appears to be the most important phase in the lead up to successful mating as it is a prelude to a chain-sequence of events that follow. Auditory and visual stimuli are used in evoking sexual responses culminating in a crouching response of a sexually receptive female. In the ostrich, a specific display called 'kantling' (Fig. 3.1a) may provide a stimulus for the female to crouch (Bertram 1992; Bubier et al. 1998; Rozenboim et al. 2003; Malecki and Martin 2005). But if kantling is not displayed the male approaches the female with wings raised in the air while making short and quick steps ('dancing on tip toes' – Bubier et al. 1998) stimulating the female to crouch (Fig. 3.1b). In the emu, on the other hand, the male stimulates the female by emitting grunting sounds while elevating his front feathers and curving his neck (Fig. 3.1c). Then he approaches the female and follows her closely behind (Fig. 3.1d) until she assumes crouching position (Malecki et al. 1997a).

Further behaviour of the receptive female is elicited by tactile stimulation from the male during mounting. In the emu, such stimulation is provided when a male

Fig. 3.1 (a) Kantling to a human (photo – P.K. Rybnik Trzaskowska); (b) 'dancing on tip toes' (photo – C.K. Cornwallis); (c) male emu in a courtship displays (photo – I.A. Malecki; (d) male emu following the female that is about to crouch (photo – I.A. Malecki)

touches the female's back with his chest, while, in the ostrich it occurs when the male places his chest and the right leg on the female's back (Rybnik et al. 2007). In response to pressure on the female's back, the ostrich female raises her head to its full height, or she places it low, just above or on the ground while clapping her beak. The female emu, on the other hand, curves the neck while suspending it above the ground (Malecki and Martin 2004). Tactile stimulation also elicits elevation of the tail and relaxation of the cloaca. In the emu, the male stimulates the female by touching her cloacal opening with his partially protruding phallus while he intends to gain intromission. In the ostrich, the male repeatedly strokes the female's cloaca with a protruding phallus before gaining intromission. A stimulated female emu or ostrich may bring the vaginal orifice close to the vent but it has not been seen to be protruding past the vent lips.

During intromission and ejaculation, the female remains on the ground under the male's weight but if she feels threatened she may terminate crouching and get up leaving the male, who falls off her. The male emu grabs the neck skin with his beak while ejaculating (Malecki et al. 1997a), behaviour similar to that observed in the domestic chicken or duck (Wood-Gush 1957; Kruijt 1964; Tan 1980). The male ostrich, on the other hand, arches his neck and swings it left to right grunting at the same time (Rybnik et al. 2007). The stimulatory role of these behaviours remains uncertain.

A dismount and post-coital display are associated with termination of female receptivity and refractoriness in the male. Preening feathers jointly rather than chasing each other around the paddock may be interpreted as successful copulation that may serve to reinforce the bond.

3.2.1.2 Stimuli Evoking Sexual Responses in a Male

In both emus and ostriches courtship behaviour and crouching have a stimulatory role for the male. In the emu, a receptive female curves her neck, elevates neck feathers and emits low frequency booming sounds. If the male responds and courts the female back, she would solicit by raising the tail and when approached, walk away keeping her front low and tail high (Fig. 3.2a) and then she would crouch (Fig. 3.2b). The female ostrich may signal her receptivity by solicitation display (Fig. 3.2c) that is characterised by lowering the neck, flapping her lowered wings and clapping the beak (Bubier et al. 1998). Crouching displays may then follow (Fig. 3.2d).

The strongest single stimulus that can elicit a sexual response in a male is a crouching female. As in turkeys (Shein and Hale 1974), chickens (Wood-Gush 1957; Kruijt 1964) and ducks (Tan 1980), the strength of this stimulus plays a key role in developing semen collection methods for emus and ostriches, although response to a crouch varies with the male. Apart from a response to a sexual partner, the male emu and ostrich also take crouching as an opportunity to copulate, because they make copulation attempts at sitting females who are not their mates or did not sit for them (Bubier et al. 1998; Malecki et al. 1997a; Rybnik et al. 2007).

3.2.2 Natural Mating, Normal Sexual Behaviour and Welfare

Whether in pairs (emu, ostrich), trios and quads (ostrich) or in colonies (emu, ostrich) welfare problems related to freedom of expression of normal sexual behaviour may occur on farms and sometimes seem unavoidable. Farming practices may interfere with natural selection of sexual partners when farmers pair birds of their choice. If individuals of artificial pairing do not have a good relationship, fighting may occur, resulting in physical harm. Also, the bond of a successful pair may become affected during the breeding season or towards the end of the season, when males or females are captured and temporarily or permanently separated. This is practised on some farms during the season to facilitate rotational mating. In the ostrich, separation can reduce the egg output and egg fertility (Cloete et al. 1998; Malecki et al. 2004). In the emu, this may not be such a problem provided the male remains in close and visual proximity of the female (Malecki and Martin 2002a, b).

Fig. 3.2 (**a**) Female emu in solicitation display (photo – I.A. Malecki); (**b**) crouching female emu (photo – I.A. Malecki); (**c**) female ostrich in solicitation display (photo – P.K. Rybnik-Trzaskowska); (**d**) crouching female ostrich – (photo – P.K. Rybnik-Trzaskowska).

At the end of the season, separation is carried out to terminate laying to give birds a rest so they regain body condition for the next season. This practice may be stressful. In the short-term, interference with pair bonding, natural mating and free expression of sexual behaviour is not desired. It may lead to inhibitions of normal sexual responses, inhibition of ovulation, poor sperm retention and lowered libido. The long-term consequence of such interference is not known.

3.2.3 Natural Mating, Abnormal Sexual Behaviour and Welfare Implications

When both males and females express what are considered to be sexual approaches towards humans this behaviour is considered abnormal. Captive breeding of ratites is affecting their early experience. The effect of this experience is observed by following up responses of the chicks and juveniles. The expression of sexual behaviour by the adult birds towards humans is seen in females as crouching and in males as attempting to mate. It is likely that a long period of rearing from hatching to puberty (up to 24 months) affects sexual choices birds make when they are ready to reproduce. When birds are hatched by providing them artificial shelter and giving food and water daily, emu and ostrich chicks ‘bond’ with their caretakers, the clothes and colours worn and the noises they make. Some birds (that subsequently express sexual behaviour towards humans) may not pair successfully with their conspecifics. The lack of normal sexual interactions between imprinted and non-imprinted birds may lead to agonistic encounters, removal from the enclosure or death of one individual. This needs to be addressed in future studies because while on one hand, ‘imprinted’ birds may be unsuccessful breeders in natural mating system and suffer as a consequence, on the other hand they may be successful under the artificial insemination system. Likewise we should determine to what extent imprinting affects temperament because imprinted birds seem tamer, better adapted and easier to farm than non-imprinted birds.

3.3 Semen Collection

The size of ratite species, its unique sexual behaviour, or the quality of collected ejaculates prompted the development of methods that minimise the impact on welfare of the ratite male such as the manual massage. Ratites are large birds and difficult to restrain. Even birds that do not show fear of humans or appear friendly become stressed if physically restrained. Therefore, the semen collection methods need to be based on a good understanding of sexual behaviour, where appropriate stimuli that evoke normal sexual responses or prevent their inhibition are applied.

3.3.1 Massage Stimulation

Von Rautenfeld (1977) was the first to propose a manual massage method for the ostrich. This method is most commonly used in poultry (Lake and Stewart 1978). The difference in the approach in these two vastly different groups of birds is that the massage technique in poultry is applied to the lumbar and abdominal regions, while in the ostrich it is applied predominantly on the phallus. However, the problem is that the male ostrich needs to be caught, restrained and hooded by a few people, which causes considerable stress to the male. Then the male ostrich is confined and physically restrained in a specially constructed ‘crush’ in order to prevent injury to the bird and the handlers (Fig. 3.3). The phallus is then extruded out of the cloaca and massaged. Bertschinger et al. (1992), Irons et al. (1996), Hemberger et al. (2001) and Lambrechts (2004) used this method but, despite some improvements they made, this method could not guarantee quality ejaculates

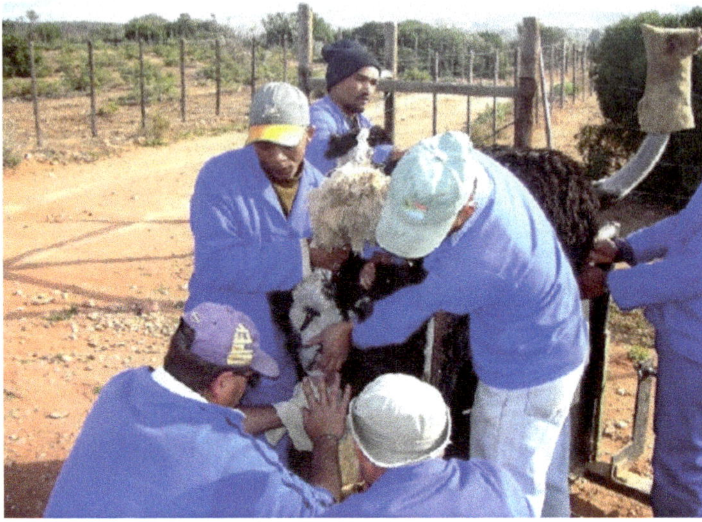

Fig. 3.3 Collection of semen from the ostrich by manual massage (photo: I.A. Malecki)

routinely and stress to the birds remained high. They found this approach labour intensive and unreliable, and as a consequence the ostrich farming and research institutions did not adopt it.

The massage method was also used in the cassowary (Gee et al. 2004) but again the male had to be physically restrained by a few people, and the resulting stress (and position of the bird) could not guarantee a desired outcome nor that it could be used on a regular basis.

3.3.2 Female Stimulus in Semen Collection (Teaser Methods)

Using the innate sexual behaviour of emus and ostriches, two methods for semen collection into an artificial cloaca (AC) have been developed. One approach is based on sexual responses directed towards conspecifics, in which the female is a stimulus for the male. The other approach is based on sexual behaviour directed towards humans so that the human is a stimulus.

In the emu, (Malecki 1993; Malecki et al. 1997a) and ostrich (Rozenboim et al. 2003; Ya-jie et al. 2001; Rybnik et al. 2007), the teaser method follows the normal sequence of courtship and mating behaviour. Usually, docile pairs that co-operate with the handlers are used and ejaculates are collected by interrupting mating. While early developers of the ostrich teaser method used docile pairs, Rybnik et al. (2007) also trained males that were aggressive. The initial inhibition of male responses to the crouching female (probably due to fear of a human standing nearby) can be overcome by adjusting the distance between the male and a human. Later, a human becomes a stimulus.

3.3.2.1 Emu Teaser Method

As soon as the collector appears, the male emu assumes courting display, willingly walks out of his enclosure, enters the female's enclosure and he follows her in a courtship display. The male stimulates the female by emitting grunting sounds, puffing his neck feathers and following her closely behind. Upon female submission and crouching, the male performs mounting (Malecki et al. 1997a) while the female lowers her neck and relaxes the cloaca. When the male moves his phallus close to the female cloaca, it is directed into the artificial cloaca (AC) and the male is induced to ejaculate (Fig. 3.4a). The male then dismounts, preens and walks besides the female, feeding or inspecting the area, and then he is directed to walk back to his enclosure.

3.3.2.2 Ostrich Teaser Method

Soon after a collector appears near the holding yard, the male observes the collector and the female teaser and follows the collector walking towards the teaser. As the collector comes near the female, she assumes the crouching position, and upon stimulation by crouching, the male attempts to mount the female. The collector then approaches the birds from behind (Fig. 3.4b) and when the male begins stroking the female's cloaca with his phallus, the collector places the AC by the female's side, and directs the phallus into the AC. At the beginning, this stage may be difficult for the male and the collector but stimulation by the AC overcomes the problem. When ejaculation commences into the AC the male performs the same repertoire of behaviours that are observed during copulation with the female, arching his neck and swinging his head sideways and grunting.

3.3.2.3 The Artificial Cloaca as a Stimulus

The AC is made to mimic the vagina/cloaca of the female; the same principle that is used in semen collection for livestock animals. The inner space between the PVC case and rubber liner is filled with warm water at the bird body temperature (Malecki et al. 1997a, Rybnik et al. 2007). A cold cloaca will not evoke a rapid ejaculation response. It can delay the response or may even become inhibitory and discourage the male.

Fig. 3.4 (**a**) Collection of semen from the emu using the teaser method (photo: I.A. Malecki); (**b**) collection of semen from the ostrich using the teaser method (P.K. Rybnik-Trzaskowska); (**c**) collection of semen from the emu using the non-teaser method (photo: I.A. Malecki); (**d**) collection of semen from the ostrich using the dummy (non-teaser) method (photo: P.K. Rybnik-Trzaskowska)

3.3.2.4 Selection of Teaser Females

In the emu and ostrich, a teaser female plays a critical role in training, collection success, ejaculate repeatability and maintenance of male libido. Therefore it is very important to select a good teaser. There are few studies of this and selection criteria and characteristics of good teaser females are limited (Malecki et al. 1997a; Rozenboim et al. 2003; Rybnik et al. 2007). The initial selection is based on readiness to sit for a human and the subsequent crouching display so that the teaser female evokes good responses from a male. This is particularly important for males with a shy temperament that take relatively long to respond. Subsequently, the teaser needs to remain in a crouching position until ejaculation is complete. Selection of the teaser is complete after testing the female's response to a human and then to a male. In the emu, the size and temperament of the females requires particular consideration because, in this species, the female is generally larger than a male and she appears to play a dominant role in the natural mating system (Blache et al. 2000). Some teasers can show preferences for particular males – this is unavoidable; so several teaser females are needed and used interchangeably when required. It is yet to be demonstrated that separation of the male from the female, as used in the emu, is feasible in the ostrich. For ostriches, additional female criteria may need to be identified.

3.3.3 *Human Stimulus in Semen Collection (Non-Teaser Methods)*

In the non-teaser method, males are trained by taking advantage of courtship behaviour directed towards humans. Such courtship has been described in emus (Malecki et al. 1997a) and ostriches (Bubier et al. 1998; Rozenboim et al. 2003; Rybnik et al. 2007). This human–bird interaction is desirable for training as it leads to expression of the male sexual behaviour. The male courtship is then returned by giving him an opportunity for further expression to the point of ejaculation. In the case of the emu (Malecki et al. 1997a), the male mounts the arm and shoulder of the semen collector, whereas in the case of the ostrich, the male mounts a dummy (Malecki and Martin 2005; Rybnik et al. 2007). The difference in the choice of the mounting object is due to differences in the mounting behaviour between the two species, body size of the male and potential risk of any agonistic interactions that may occur between the male and the human inside the male enclosure. The male emu mounts the female with both of his legs on the ground. There is thus little weight that the human needs to take on the arm when the male emu rests his body during ejaculation. The male ostrich rests his right leg and most of his body on the female's back. Such weight is a risk that a human cannot take so a dummy is used. Temperature of the AC is as important as it is for the teaser method.

3.3.3.1 Emu Non-Teaser (Human) Method

Semen collection is carried out with the collector inside the male's enclosure. The male emu, seeing the collector approaching, begins courtship by curving his neck and elevating neck feathers, grunting sometimes and approaching the collector. Then either the collector first squats in front of the male or the male squats first in front of the semen collector. While moving on his hocks forward the male attempts to mount the collector's arm. The male may be further stimulated by rubbing his abdomen with a hand while holding the AC directed towards the protruding phallus. Placing the AC on the phallus induces ejaculation. During ejaculation the male grabs the collector's jacket with his beak and rests some of his body on the collector's arm (Fig. 3.4c).

3.3.3.2 Ostrich Non-Teaser (Dummy) Method

When the collector approaches the male enclosure with the dummy, the male may already be kantling or standing by the gate waiting for the dummy to be placed in front of him. Once the dummy is placed, the ostrich male comes up with his wings lifted in the air and attempts to mount the dummy. The male puts his right leg on the dummy, spreading and waving his wings and swinging the head from side to side. When intromission is achieved in the AC the male ejaculates while arching his neck forward and swinging his head sideways and grunting (Fig. 3.4d), a sequence of behaviours performed during normal copulation or during a teaser collection method. Once ejaculation is complete the male dismounts and walks away (Rybnik et al. 2007).

3.3.3.3 Semen Collection Methods and Bird Welfare

The welfare of ratite males trained to the teaser and non-teaser methods do not appear to be compromised because normal sexual responses are evoked and the males co-operate well. The fact that the methods are reliable and guarantee physiological ejaculates on regular basis suggests ejaculates are not obtained under stress. Given that the new methods: (1) require one person to train the birds and to collect semen routinely; (2) are relatively stress free and physically not very demanding; they are not only animal but also human friendly.

One of the aspects that may affect male welfare is semen collection frequency. Since the methods for ratites are based on voluntary ejaculation it may be assumed a male may refuse it at any time. However, some high libido males may be driven to a high number of ejaculations if given an opportunity. This may lead to male exhaustion, loss of libido or inhibition. The optimum collection frequency has been determined for emus trained to both methods (Malecki et al. 1997b) and for ostriches trained to the dummy method (Bonato et al. 2011). They showed that ejaculations of up to twice a day do not adversely affect male libido or ejaculate quality. A similar response of ostrich males to collection frequency using a teaser method is expected. Since males vary in libido the frequency of collection needs to suit the male. If higher than optimal frequency is attempted but not withdrawn on time inhibitions will develop. Reversibility of libido loss has not yet been studied.

In respect to the welfare of the female, mounting is only feasible if a female crouches sexually. Hence, as is in the case of the male emu or ostrich, the female appears to be also in control of her welfare as she may refuse to crouch. In the case of the ostrich, the female may, however, be subjected to aggression from the male during arranged joining for semen collection. The behaviour of both birds needs to be assessed and if the risk is anticipated the birds should not be put together or they should be separated as soon as a risk of injury to a female arises. The reverse may occur in the emu, unless a male can dominate the female. Although the teaser female crouches for a human it does not mean she will do the same for the male.

Thus, testing female receptivity by inducing crouching for a human is not very predictive of a sexual crouch for the male.

The fact that Rozenboim et al. (2003) found the artificial vagina (AV) stressful for their pairs may be explained by the complicated approach of putting a rubber on the phallus that (at this stage) did not resemble the female cloaca. Hence inhibitions due to phallus manipulation and delay with intromission could develop.

3.4 Artificial Insemination

Attempts at using natural mating to proliferate desirable genes are not very successful and cannot be compared with artificial insemination because of the individual preferences for chosen mates that both emus and ostriches express. Moreover, the use of artificial insemination greatly extends the number of females each male can fertilise (Malecki et al. 2006). In poultry, artificial insemination is a routine procedure that uses restraint, but females appear responsive to the insemination technique and the size of the female does not pose a problem for the handlers. Female ratites, however, due to their size, need a different approach because physical restraining and the insemination procedure that usually follows may be stressful. For the emu a behavioural approach has been developed (Malecki and Martin 2004), but female ostriches, until recently (Malecki et al. 2008), have had to be caught and restrained for this procedure (Von Rautenfeld 1977).

3.4.1 Use of Stimuli to Evoke Sexual Responses in the Female and Female Receptivity

Since the discovery of female crouching in response to stimulation by the handler, a human approach has been used to induce female emus and ostriches to crouch. Females that crouch can potentially be inseminated if they are further stimulated and the insemination procedure is carried out only with the compliance of the female (Fig. 3.5a, b). Having been able to evoke the crouching response in the female, a human can evoke further responses in a female that would normally be induced by the male. Touching and rubbing the female's back, base of the tail and sides of her abdomen may cause the female to lower her body and neck towards the ground, elevate her tail, expose the vent and relax the cloaca. In the ostrich, a hand may then be introduced into the proctodeum and the fingers directed to the left through the urodeum and to the oviduct opening, where the insemination straw is then guided. In the female emu, a relaxed cloaca would allow insertion of a speculum and the insemination straw. Following removal of the straw, the female may be further stimulated by pressing her back with a hand or by just placing a hand on her back, rubbing the sides and the neck. This point may be critical to female

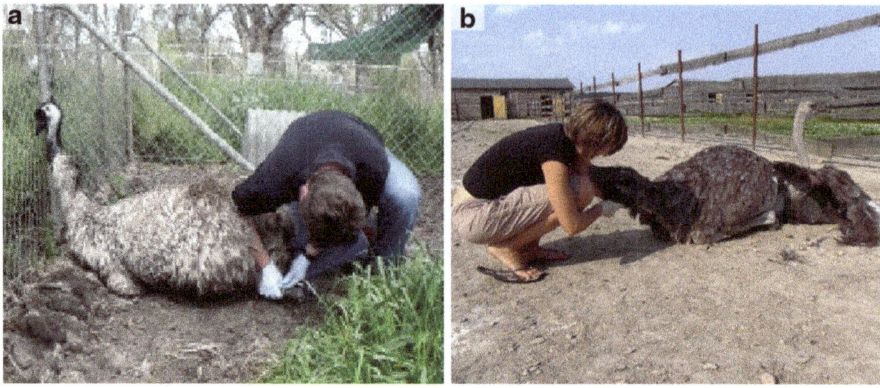

Fig. 3.5 (**a**) Artificial insemination of a female emu following sexual crouch (photo – I.A. Malecki); (**b**) artificial insemination of a female ostrich following sexual crouch (photo – P.K. Rybnik-Trzaskowska)

receptivity and even subsequent ovulations, because if the insertion of the straw is not thorough and not carried out at the optimal time, it is likely to adversely affect the female. The fact that the vagina of the female ostrich appears more convoluted and twisted than that of the female emu suggests greater care needs to be taken to inseminate the female ostrich than the emu.

Although, crouching behaviour has been demonstrated in both species (Malecki et al. 1997a, 2008; Rybnik et al. 2007), its relationship to female receptivity is unclear. If we accept that only sexually receptive females crouch (Shein and Hale 1974; Tan 1980), non-receptive females cannot be inseminated on their own accord, hence insemination of sexually crouching females may be regarded as female friendly. The crouching behaviour has a seasonal pattern. The female is mainly observed crouching in the breeding season and remains at distance with a human outside the breeding season. She is less willing to crouch or cannot be induced to crouch when followed and sometimes she is aggressive. Again, crouching thus appears sexually motivated, given that sex hormones are needed in the blood plasma for expression of this behaviour. Apart from seasonal variation in female receptivity, there may still be a diurnal pattern related to ovulation–oviposition cycle due to variations in the concentration of the hormones associated with ovulation and oviposition that change every 48 h in the female ostrich (Bronneberg et al. 2005) and every 72 h in the female emu (Malecki and Martin 2002b).

Termination of receptivity does not appear to be caused by direct palpation of the cloaca, vaginal orifice or straw insertion in the vagina. Receptive females, especially in emus rather than ostriches, tend to remain on the ground after insemination. However, if the insemination procedure causes stress to the female she may terminate her receptivity. During natural mating, termination of receptivity is more apparent as the female emu or ostrich gets up soon after a male dismounts from her and she is not willing to immediately crouch again.

Usually females that lay or are about-to-lay eggs crouch sexually, there are females that do not lay eggs or have never laid an egg but also crouch. Such females may not be as responsive to stimulation as females that lay. It is important to determine if crouching is sexual in nature and has a pattern related to sex hormones concentrations in blood plasma or if it is a learnt behaviour originating from non-sexual submissive response. Understanding of the relationship between crouching and the ovulation–oviposition cycle will assist in better defining the optimal insemination time and the time when the female is most receptive. It may improve the insemination procedure and contribute to better female welfare.

3.4.2 Artificial Insemination and Female Welfare

Since female emus and ostriches develop a responsive behaviour and crouch they allow insemination to be carried out without restraint and their welfare is not compromised (Malecki and Martin 2004; Malecki et al. 2008). After the female crouches, stimulation is applied to her back and around the base of the tail. The female responds and no restraint or any force is required. Therefore up to this point the procedure appears to be welfare neutral to the female.

Semen is deposited into the vagina following insertion of the inseminating straw and if insertion of the straw is thorough and carried out at the optimal time, it is not likely to adversely affect the female. If careful attention is paid and the inseminator works with the female as she contracts and relaxes vagino-cloacal muscles, such an insemination procedure does not appear to upset the female. There are also some subtle between female differences in receptivity pattern that should be realised and remembered so the insemination procedure should be tailored to individual females. Some females require constant pressure on their back during insemination so assistance from a second person may be needed.

Artificial insemination can also be carried out on some non-crouching female ostriches. Physical restraint is not needed but the female needs to show sufficient arousal. Female ostriches may stand for a human and may not even need a hood. But if the female does not respond to stimulation and their cloaca is not relaxed, insemination should not be performed, as it may be stressful and lead to the development of inhibition as the procedure becomes invasive. Sexual crouching and arousal of the female is required to avoid stress and this can be achieved if insemination is carried out at the optimum time and following sufficient stimulation. Female emus can rarely be inseminated with a speculum when standing.

The female that continues displaying receptivity after inseminations and continues producing eggs indicates that her welfare has not been grossly affected by the insemination procedure. In addition, a thorough and timely insemination technique will ensure most semen being retained in the oviduct for storage and fertilization of subsequent ova. This would maximise the efficient use of the semen; however, we are yet to fully evaluate all procedures. Hence further studies of the emu and ostrich ovulation–oviposition cycle in relation to the timing of insemination are needed.

A time-wise insemination will not only ensure good female welfare but it will also contribute to retention of semen and better female fertility that may result in less frequent inseminations needed to maintain the female fertile.

3.4.3 *Effect of Early Experience on Female Welfare*

We have observed that early experience of humans may have at least two types of consequence for females that may be more concerning for the female ostrich than emu. One is that if the female ostrich displays to a human and crouches such a female may not be a good mate for her conspecific male. The female welfare may be affected as she may be chased, attacked, kicked out of the enclosure or even killed. The same may occur if the male shows a preference for a human. Such a male tends to be aggressive towards the female probably as a result of competition for a human. The reasons behind such interactions should be determined in future studies. The female emu, probably due to her dominance or less aggressive nature of the male, has fewer welfare implications.

The second consequence appears to have a better outcome for the female than the first. The female that displays and crouches for humans, and lays eggs, may be maintained without a male and she would not be then subjected to male harassment. Such females appear stress free when a human is around, are easier to handle and are not adversely affected by artificial insemination.

3.5 Final Considerations

Little so far has been published on the behaviour of ratites in the farming environment and on most farms there is usually little interaction between farm personnel and birds due to large flock sizes and infrequent human involvement. Humans interact less with adult birds than they do with the chicks. On ostrich farms, human–bird interactions are particularly limited or even avoided due to the aggression of ostriches. Ostriches appear to be selective and they may not be aggressive towards some people, particularly those with whom they become attached to over the first months or years of their life. With some male ostriches, a human can walk inside the enclosure when attempting to induce crouching or present a dummy to a male (Fig. 3.6a, b). Imprinting appears to play a role in this bonding process, which seems to have a few stages and/or critical periods (hatching and early rearing, juvenile about 1 year old and puberty) that future studies should address. Human selection may already be expressed during juvenile (Fig. 3.6c) or pubertal (Fig. 3.6d) stages of development. Aggression in emus or ostriches is generally expressed by birds fearful of humans. Selection for a docile temperament could eliminate aggression and create more humane and animal-friendly farming. Interestingly, little aggression is nowadays observed on emu farms without purposely

Fig. 3.6 (**a**) Semen collector inside the birds' enclosure walking past the front of the male ostrich (photo – I.A. Malecki); (**b**) semen collector inside the birds' presenting a dummy to the male ostrich (photo – I.A. Malecki); (**c**) a juvenile male ostrich in a kantling display to a person standing by the pole fence (photo – P.K. Rybnik-Trzaskowska); (**d**) juvenile female ostriches being examined by a person to whom they crouched spontaneously (photo – P.K. Rybnik-Trzaskowska)

selecting for it, probably as a result of eliminating less adaptable individuals and epigenetic generational changes. Ostrich farming appears to be undergoing this process, but to accelerate it genetic selection needs to be applied and special husbandry practices need to be developed. Interest in humans, lack of fear of humans and expression of sexual behaviour in the presence of humans or directed towards humans are indicative of changes in the social and sexual behaviour of emus and ostriches that are being affected by farming. In poultry, a loss of pair bonding and spontaneous egg laying are a consequence of domestication (Craig 1981; Sossinka 1982). The domestication process appears to have a similar effect on emus and ostriches. Friendly behaviours towards human (approach, following, solicitation, kantling, eating from a hand, no fear of touch or rubbing the back, neck or tail and others) and responses to sexual stimuli that the females and males display might result from imprinting to humans (Lorenz 1937), a phenomena that needs detailed studies in ratites. Crouching for humans, spontaneous laying and the animal-friendly methods for semen collection and artificial insemination that have been developed give prospects for developing technologies adaptable by farming and conservation. Through the use of artificial insemination technology, sustainability of farming and preservation of wild populations are achievable; however, our knowledge of ratite physiology and behaviour needs to be expanded so we can ensure their best welfare while ratites continue to evolve in natural and farming environments.

Acknowledgements We thank the Rural Industries Research and Development Corporation (Australia), the Western Cape Institute of Agriculture (South Africa); the Polish Committee for Scientific Research (Ministry for Education, Poland) and the Japanese Society for the Promotion of Science for support. Dr P.K. Rybnik-Trzaskowska was supported by the Australia Endeavour Research Fellowship when working on this chapter.

References

Bertram BCR (1979) Breeding system and strategies of ostriches. In: Proceedings of the 17th International Ornithology Congress, Germany, pp 890–894. Deutsche Ornithologen-Gesellschaft, Berlin

Bertram BCR (1992) The ostrich communal nesting system. Princeton University Press, Princeton, New Jersey

Bertschinger HJ, Burger WP, Soley, JT, de Lange JH (1992) Semen collection and evaluation of the male ostrich. In: Proceeding of the Biennial Congress of the South African Veterinary Association, 7–10 September 1992, Grahamstown, South Africa, pp 154–158

Blache D, Barrett CD, Martin GB (2000) Social mating system and sexual behaviour in captive Emus Dromaius novaehollandiae. Emu 100:161–168

Bonato M, Rybnik PK, Malecki IA, Cornwallis CK, Cloete SWP (2011) Twice daily collection yields greater semen output and does not affect male libido in the ostrich. Anim Reprod Sci 123:258–264

Bronneberg RGG, Taverne MAM, Dieleman SJ, Decuypere E, Bruggeman V, Vernooij HCM, Stegeman AJ (2005) Ultrasonographic observations of the oviduct and plasma progesterone, estradiol and LH profiles during the egg laying cycle in female ostriches. In: Proceedings of the 3rd International Ratite Science Symposium, Carbajo E (ed) 14–16 October 2005, Madrid, Spain, pp 91–95. Imprenta, Madrid, Spain

Bubier NE, Paxton CGM, Bowers P, Deeming DC (1998) Courtship behaviour of ostriches (Struthio camelus) towards humans under farming conditions in Britain. Br Poult Sci 39:477–481

Cloete SWP, Van Schalkwyk SJ, Brand Z (1998) Ostrich breeding – Progress towards a scientifically based strategy. In: Ratites in a Competitive World, pp 55–62. Huchzermeyer FW (ed.) Proceedings of the 2nd International Scientific Ratite Conference, Oudsthoorn, South Africa, 21–25, September 1998 De Jongh's Printers – Printers and Designers, Strand, South Africa

Coddington CL, Cockburn A (1995) The mating system of free-living Emus. Aust J Zool 43: 365–372

Craig JV (1981) Domestic animal behavior: causes and implications for animal care and management. Prentice-Hall Inc, Englewood Cliffs, NJ

Gee GF, Bertschinger H, Donoghue AM, Blanco J, Soley J (2004) Reproduction in nondomestic birds: physiology, semen collection, artificial insemination and cryopreservation. Avian Poult Biol Rev 15:47–101

Hemberger MY, Hospes R, Bostedt H (2001) Semen collection, examination and spermiogram in ostriches. Reprod Domest Anim 36:241–243

Irons PC, Bertshinger HJ, Soley JT, Burger WP (1996) Semen collection and evaluation in the ostrich. In: Deeming DC (ed) Improving our Understanding of Ratites in a Farming Environment. Ratite Conference, Oxfordshire, pp 157–159

Kruijt JP (1964) Ontogeny of social behaviour in Burmese Red Jungle fowl (Gallus gallus spodiceus) Bonaterre. Behaviour 12(S):1–201

Lake PE, Stewart JM (1978) Artificial insemination in poultry, ministry of agriculture fisheries and food, bulletin 213. Her Majesty's Stationery Office, London

Lambrechts H (2004) Reproduction efficiency of ostriches in commercial farming systems. PhD Thesis, University of the Free State, Bloemfontein, South Africa

Lorenz KZ (1937) The companion in the bird's world. Auk 54:245–273

Malecki IA (1993) Sexual behaviour and semen collection in the Emu (Dromaius novaehollandiae). P.G. Dipl. Thesis, The University of Western Australia, Perth

Malecki IA, Martin GB (2002a) Fertility of male and female emus (*Dromaius novaeholandiae*) as determined by spermatozoa trapped in eggs. Reprod Fertil Dev 14:495–502

Malecki IA, Martin GB (2002b) Fertile period and clutch size in the emu (*Dromaius novaeholandiae*). Emu 102:165–170

Malecki IA, Martin GB (2003) Sperm supply and egg fertilisation in the ostrich (Struthio camelus). Reprod Domestic Anim 38:429–435

Malecki IA and Martin GB (2004) Artificial Insemination in the Emu (Dromaius novaehollandiae): Effects of numbers of spermatozoa and time of insemination on the duration of the fertile period. Animal Science Papers and Reports, 3, 315–323

Malecki IA, Martin GB (2005) Development of reproductive technology and fertility assessment for the emu and ostrich farming. Rural Industries Research and Development Corporation, Australia, RIRDC Publication No 05/200

Malecki IA, Martin GB, Lindsay DR (1997a) Semen production by the male Emu (Dromaius novaehollandiae). 1. Methods for semen collection. Poult Sci 76:615–621

Malecki IA, Martin GB, Lindsay DR (1997b) Semen production by the male Emu (Dromaius novaehollandiae). 2. Effect of collection frequency on the production of semen and spermatozoa. Poult Sci 76:622–626

Malecki IA, Cloete SWP, Gertenbach WD, Martin GB (2004) Sperm storage and duration of fertility in female ostriches (Struthio camelus). S Afr J Anim Sci 34:158–165

Malecki IA, Cloete SWP, Martin GB (2006) Future directions in breeding of ratites. Latin American Ratite workshop, October 2006, Sao Paulo, Brazil, Conference CD

Malecki IA, Rybnik PK, Martin GB (2008) Artificial insemination technology for ratites: a review. Aust J Exp Agric 48:1284–1292

Rozenboim I, Navot A, Snapir N, Rosenshtrauch A, El Halawani ME, Gvaryahu G, Degen A (2003) Methods for collecting semen from the ostrich (Struthio camelus) and some of its quantitative and qualitative characteristics. Br Poult Sci 44:607–611

Rybnik PK, Horbanczuk JO, Naranowicz H, Łukaszewicz E, Malecki IA (2007) Semen collection in the ostriches using a dummy or a teaser female. Br Poult Sci 48:635–643

Sauer EGF, Sauer EM (1966) Social behaviour of the South African Ostrich (Struthio camelus australis). Ostrich 6(S):183–191

Shein MW, Hale EB (1974) Stimuli eliciting sexual behaviour. In: Beach FA, Robert E (eds) Sex and behaviour. Krieger Publishing Company, New York, pp 439–482

Sossinka R (1982) Domestication in birds. In: Famer DS, King JR, Parkes KC (eds) Avian biology, vol 6. Academic, London, pp 373–404

Tan NS (1980) The frequency of semen collection and semen production in Muscovy drakes. Br Poult Sci 21:265–272

Von Rautenfeld DB (1977) Mitteilungen zur kunstlichen besamung, geschlechts und altersbestimmung beim strauʃ (Struthio camelus australis, Gurney). Der Praktische Tierarzt 5:359–366

Wood-Gush DGM (1957) Fecundity and sexual receptivity in the Brown Leghorn female. Poult Sci 37:30–33

Ya-jie JI, Yan-bo Y, Wu-zi D (2001) Studies on ostrich semen character and semen storage at low temperature. J Econ Anim 5:49–54

Chapter 4
Incubation and Chick Rearing

D.C. Deeming

Abstract Welfare considerations during incubation and for the first 3 months post-hatching are described for ratite chicks. Welfare of embryos is discussed in the context of the animal being sentient and able to elicit a change in its environment. It is suggested that during development, typical definitions of what constitutes welfare cannot be easily applied to embryos and indeed they may be inappropriate. Only after internal pipping into the air space does an embryo exhibit a demonstrable ability to elicit a response from its parents when incubation conditions are not optimal. Nevertheless, good practice will optimise the incubation environment and yield high hatchability and chick quality, which automatically serves to maximise welfare.

Compared with poultry species, our understanding of the requirements of ratite chicks is very poor. There are limited scientific studies into appropriate rearing conditions and few that are directly related to welfare. Welfare considerations during rearing are described in relation to factors such as behaviour, feeding, environment, transport and health issues. Much of this work has been carried out with ostriches, with little work on emus or rheas. Chick welfare in all commercially important ratites requires more targeted research in order to promote best practice around the world.

Keywords Embryo · Emu · Health · Incubation rearing · Ostrich · Rhea

D.C. Deeming
Department of Biological Sciences, University of Lincoln, Riseholme Park, Lincoln LN2 2LG, UK
e-mail: cdeeming@lincoln.ac.uk

P.C. Glatz et al. (eds.), *The Welfare of Farmed Ratites*, Animal Welfare 11,
DOI 10.1007/978-3-642-19297-5_4, © Springer-Verlag Berlin Heidelberg 2011

4.1 Introduction

Incubation is crucial in the process of production of avian offspring. The rearing of the hatchlings is as crucial but is often less under the control of the parents – there are many factors that can cause mortality of young birds. Human interaction with birds, be they fully domesticated poultry or waterfowl, semi-domesticated ratites, or wild birds under aviculture, has meant that reproduction is controlled by people because eggs are artificially incubated and the chicks are regularly reared under artificial conditions rather than by their parents. This means that the success of reproduction is dependent on people being able to perform the required procedures effectively. All too often artificial incubation or rearing can fail through human error or negligence and this can impact upon the welfare of the birds involved. In this chapter, the factors affecting the welfare of ratites during the incubation and rearing stages up to a chick age of 3 months of age are considered. Discussion is limited to the ostrich (*Struthio camelus*), emu (*Dromaius novaehollandidae*) and rhea (*Rhea americana*), which are the three species that have been taken into captivity for commercial exploitation and so are most likely to have their welfare compromised.

Incubation is described and the possible impact it can have on the welfare of the embryo is discussed in detail. However, most of the chapter is dedicated to the welfare of ratite chicks and the factors that can determine whether the birds being reared reach their full potential. Regrettably relatively little is known about rearing of ratite chicks, which compromises our ability to provide appropriate conditions that could serve to maximise welfare.

4.2 Incubation

Avian incubation is relatively simple process: heat is applied to the egg to raise it to a temperature suitable for embryonic development, typically 37–38°C in poultry; there is provision of an appropriate level of humidity to allow for optimal loss of weight; there are frequent changes of air around the egg to provide oxygen; and the egg is turned on a regular basis (once an hour in poultry). Birds provide these environmental conditions within a nest constructed for the purpose of holding eggs and chicks and by application of a brood patch to the egg to transfer heat energy to raise the egg temperature (Deeming 2002a). These conditions can be mimicked by placing the egg into an artificial incubator – typically a box where the air is heated to the correct temperature and systems are devised to exchange the air and to turn the eggs on a regular basis. This is a simplistic view of incubation and people find artificial incubation a lot more difficult to get right than the birds. This has not, however, prevented development of commercial operations that involve incubation of millions of poultry eggs with great success.

Artificial incubation of ratite eggs is a relatively recent development that was associated with the increasing commercialization of ostriches in particular in the latter half of the nineteenth century. Development of an artificial incubator for ostrich eggs occurred as early as 1868 (Smit 1963) although this was rather unsophisticated, which was typical for all artificial incubators at that time. By removing eggs for artificial incubation the adult birds were not required to incubate the clutch, which was of a limited size, and so egg production could be increased and the breeding season prolonged. Successful incubation for much of the history of ostrich farming was the result of trial and error. This tended to make incubation more of an art than a science and traditional techniques persisted almost to the end of the twentieth century.

It was the development of interest in ostrich farming, and then emu and rhea production, in the late 1980s and early 1990s that saw changes in the way that ratite eggs were incubated. Firstly, the numbers of people attempting to use incubators increased dramatically even though most "farmers" had almost no experience of the process. There were differing levels of incubator sophistication, which was dependent on how much money was available to purchase the equipment. The range was from low cost, small capacity, basic boxes with heaters and egg racks through converted food chiller cabinets, and finally to the larger capacity (more expensive) commercial machines adapted from the poultry industry. Some of the latter may have held hundreds of eggs but this was still a small operation compared to the large poultry operations. Inevitably the inexperience of people, coupled with ill-informed "advice" from people with no specialisation in incubation, meant that success of ratite incubation was poor (Deeming and Ar 1999). This probably contributed to the eventual decline in popularity of ratite farming in many countries around the world (Deeming 2002b). Incubation of ratite eggs is not technically difficult, but its success does require a good understanding of what environment is required to hatch out these birds.

Artificial incubation of these species reflects the impact that the large size of the eggs has on the incubation environment (Deeming and Ar 1999). Most importantly, the air temperature in the incubator is typically between 36.0 and 36.5°C, which is a compromise temperature that allows for early embryonic development whilst ensuring that the eggs do not overheat during the last third of incubation when metabolic heat production is at its greatest (Gefen and Ar 2001). It is possible to incubate ostrich eggs at higher temperatures (e.g. 37.0°C) but only at the start of single-stage ("all-in, all-out") incubation (Deeming and Ayres 1994). Compared to poultry incubators, humidity settings for ratite eggs tend to be low (25–35% RH), which pose problems with achieving sufficient loss of water during incubation. Failure to optimise weight loss can affect the water content of the hatchling, which determines the position of the pip hole and can ultimately dictate whether a chick hatches or not (Deeming 1995a). Provision of oxygen in commercial incubators has not been examined for ratite eggs and most hatcheries adopt rates of machine aeration comparable to poultry eggs. Variation in the method of egg turning have only been examined for ostrich eggs and the position of the egg in the incubator and the angle of turn are both crucial in maximising hatchability (van Schalkwyk et al. 2000).

Other reviews of the incubation requirements of ratites eggs are available (Deeming and Ar 1999; Deeming 1993, 1997), but in this chapter the impact of incubation on the welfare of the chick is considered.

4.2.1 Incubation and Embryo Welfare

It is interesting to speculate whether the inability of people to successfully hatch ostrich, emu and rhea eggs compromises the welfare of those embryos. This will depend on the view taken about whether the welfare of embryos is a consideration in the first instance. Such a debate has not been undertaken despite the fact that this is an important topic that needs careful philosophical consideration. Here an embryologist's view is presented. Welfare consideration need to take into account how the animal perceives and interacts with its environment, and how human activity interferes with this.

Bird embryos are not free-living organisms that have independent control over their interaction with the environment. For almost all of its existence an embryo is so dependent on the life support system provided by its extra-embryonic membranes *in ovo* and the external incubation environment supplied by its parent or an incubator. This means that it is impossible for it to survive outside of the egg. Indeed, it could be argued that it is only after internal pipping, when the embryo is breathing air and it has nearly retracted its residual yolk sac, thereby allowing it proceed to break the shell and hatch, that the bird is able to survive outside the egg. Certainly, we know that embryos forcibly removed from the egg before this time cannot survive. The question arises of when embryos should be considered sentient, aware of their external environment, and able to respond to adverse incubation conditions.

Although not studied in ratites, development of the senses in the fowl embryo depends on the sense involved. Hence, auditory sensitivity starts around day 12–14 of development (out of 21) compared with visual sensitivity starting at day 18 and improving thereafter (Rogers 1995). In addition, electrical activity in the brain is first registered on day 16 of incubation (Rogers 1995). Muscular activity is initially considered passive but becomes active during the second trimester of development although control is autonomic (Rogers 1995). When, or if, embryos develop cognitive abilities has not been investigated despite their importance in development of welfare programmes. Moreover, it is difficult to be certain that a response by an embryo to any external stimulus is comparable to behavioural or physiological responses in free-living birds – it may be an autonomic reflex.

Even if the embryo perceives its environment is not optimal it lacks any mechanism to elicit a response from the incubating parent that could correct the problem. Investigation of how an embryo could influence its parents' behaviour is rare. It has been shown that hypoxia stimulates the embryo to release nitric oxide, which diffuses across the eggshell [p. 159 in Deeming (2002a)] but whether this elicits any response in the incubating adult has not been demonstrated. Therefore, for much of its life *in ovo* an embryo lacks the sensory ability to perceive its environment and any demonstrable mechanism to actively change any adverse environmental conditions.

This changes when, around 36–48 h before hatching, the embryo internally pips into the air space and is able to breathe air. Embryos can now interact with their environment by producing sounds that can elicit a response from the incubating parents. It has been shown in several species that there is increased vocalisation in response to changes in the environment (e.g. reduced temperature) that elicits a response from the adult birds (Brua 2002). This ability seems limited to the last 1–2 days of development in the shell (Brua 2002). To date, of ratites only rhea embryos have been shown to communicate with each other during hatching, which can lead to development being accelerated in younger embryos (Bruning 1973), but these observations have not been repeated.

The problem underlying the consideration of the welfare of avian embryos is, therefore, centred on the timing of the onset of sentience during development and the animal's ability to elicit a change in its environment. Embryos are not free-living individuals and for much of their life *in ovo* lack the sensory apparatus, cognitive ability and behavioural or physiological mechanisms to regulate their incubation conditions. Hence, prior to internal pipping embryos have to simply endure sub-optimal incubation conditions. Only during the hatching process are welfare considerations that typically apply to free-living individuals of relevance to embryos. Against such a background, considering the welfare of embryos prior to internal pipping may be largely of academic interest. The author believes that the ability of the organism to be free-living and to actively effect a change in its environment is a crucial argument when it comes to considerations of welfare of avian (and reptilian) embryos prior to internal pipping. Moreover, adoption of the idea by licensing authorities charged with regulating animal experimentation, that a bird embryo is sentient around 50% of its incubation period requires experimental evidence and is certainly not universally applicable. Differences in incubation period and rates of development mean that the stage of development at half way through incubation will be significantly different between species. Therefore, whilst the author does not condone deliberate abuse of embryos through use of incorrect incubation conditions, the criteria usually used to define welfare considerations in free-living animals cannot strictly apply to embryos.

There is no doubt that poor management of the eggs, and the environmental conditions during incubation, can severely impair the ability of an embryo to survive through to hatching and it often dies *in ovo*. This is undesirable on two levels. Firstly, poor hatchability is commercially important and incubation practice should aim to maximise production of the first-grade, healthy chicks. Secondly, in free-living organisms failure to provide the correct environmental conditions for an animal can be considered as a failure to consider the welfare of the individual. Logically poor incubation conditions should adversely impact upon the welfare of embryos, irrespective of whether they are equivalent to free-living organisms. It is fortunate then that providing the correct incubation conditions will automatically optimise the environment for the embryo and so should not impact upon its welfare.

Whatever the debate about the sentient ability of embryos and their opportunities to alter their environment, during commercial incubation, a goal of maximising hatchability and chick quality ultimately serves to provide the ideal incubation environment and so, by default, should maximise the welfare of the embryo.

4.3 Hatching

Any bird embryo emerges from its egg over an extended period of time that involves distinct phases. There is an initial rotation within the egg to position the body and head in the optimal orientation for the mechanical process of hatching. There is then the process of internal pipping, which involves the embryo gaining access to the air within the air space enclosed within the broad end of the egg. Thereafter there is an extended period of time (up to 24 h), which allows the embryo to inflate its air sacs and lungs and so develop a functional respiratory system. Once this is achieved the embryo proceeds to break the shell – external pipping. Again there may be a period of rest (variable in duration) whilst the bird begins to respire atmospheric air. Hatching proceeds by the embryo breaking the shell with its beak whilst rotating within the egg. This effectively enlarges the hold in the eggshell and once of sufficient size, the shell can be pushed apart by the embryo as it emerges as a chick.

The sequence of hatching for ostrich embryos differs slightly from that typically seen in other birds (Deeming 1995a). Notably internal pipping involves the beak being rubbed against the shell membranes to create a hole and then the air space is distorted to pull the inner shell membrane over the beak. After external pipping, the hard brittle shell is broken with the beak and foot and the chick emerges after rotation of around 90°, which is less than many species, e.g. the domestic fowl. Although not described in such detail hatching in the emu and rhea seems to be comparable, although it is not known whether the mode of internal pipping is the same.

The period of hatching is a critical period for the welfare of the chick. During the boom in ostrich farming in the 1990s there was a general ignorance of the process of hatching and problems arose with the management of hatching eggs. The biggest problem was the practice of assisted hatching. At worst this involved the shell being broken before internal pipping. More typically, impatience with the progress of hatching drove some farmers to crack open the air space of eggs that when candled were showing movement in the air space; they, incorrectly, assumed that the "chick would suffocate". If the chick was not forcibly removed from the shell, this did not necessarily prove fatal, but it did increase rates of water loss from the egg, which would have increased the risk of dehydration of the chick. Even though cracking the prematurely exposed bird to atmospheric air it would have had no effect on the eventual timing of hatching. So cracking an unpipped egg is unnecessary and exposes the chick to risks, making this practice of welfare concern. Notwithstanding this, assistance for chicks trapped in shells that they had broken themselves could prove useful and could save a perfectly good bird from dying because it was trapped in the shell. Another particular problem was seen in ostrich chicks that had rotated within the eggshell and away from the pip hole but without cracking the shell – these chicks appeared to suffocate (Deeming 1995a).

Although not formally surveyed, there was a general consensus in the farming community that assisted hatching did not have adverse effects. However, Deeming and Ayres (1994) showed that those chicks that were assisted to hatch were weak and grew more slowly than those chicks that hatched naturally. In general, assisted hatching can only been seen as inappropriate and having an adverse effect on chick welfare. Failure to hatch naturally can be taken to indicate that the individual bird was experiencing a problem that was affecting its ability to survive. Assisted hatch may well have temporarily prolonged the life of animals that would have normally died in the egg. One particular problem is in fungal contamination of the egg. Upon candling the author has seen ostrich embryos that had internally pipped and were active, but they died *in ovo* before external pipping; post-mortem examination revealed the air space and head of the embryo covered with fungal hyphae that would have prevented the chick from surviving even if had been helped to break the shell.

Given that during hatching the embryo is aware of its environment, is capable of free-living, and can elicit a change in its environment, the author would argue that welfare considerations of free-living ratite chicks should start within the egg at the time of internal pipping. Environmental conditions optimised during the hatching process to minimise problems. In the past, the welfare of full-term embryos has not been of the greatest importance to farmers at this time despite the fact that it is a critical time in the life of the birds. Those advising ratite farmers in the future should be mindful to emphasise this point.

4.4 Chick Quality

One concept that is common in artificial incubation of birds is that of "chick quality" yet this is very hard to define. "First grade" chicks are easy to recognise yet prove hard to define. Typically, quality is defined by the number of faults that a chick exhibits, with the best birds exhibiting no faults. All this is subjective and so attempts have been made to quantify these characteristics for poultry species (Boerjan 2002; Tona et al. 2004, 2005). The idea of measuring the length of the broiler chick to ascertain its quality (Hill 2001; Lourens et al. 2005) has yet to be fully proven to be an effective tool (Deeming 2005). Even if it is of value, as yet there seems to be no scientific explanation of the mechanism that would bring this about. The concept of chick quality in ratites has not been investigated and commercial operations have followed the lead from poultry hatcheries. Deeming (1995a) showed that low rates of weight loss during incubation adversely affected the degree of oedema exhibited by ostrich embryos, which could be considered as a reduction in chick quality.

The impact of incorrect incubation conditions has not been investigated in ratites and is only recently being considered in poultry species. The main emphasis is on maintenance of the correct incubation temperature because high temperatures (only 1–2°C above set air temperature) can have impact on chick performance post-hatching. For example, Lourens et al. (2005) have shown that rates of growth and body temperature of broiler chicks are affected by incubation temperature during the last trimester of incubation. Such effects are only now being considered and investigated in the more commercially important poultry species. Comparable investigations would be of great value for ratite species. It is anticipated that high temperatures should cause problems for ratite embryos. The larger eggs retain the metabolic heat produced by the embryo more effectively than smaller poultry eggs and so egg temperatures can rise to deleterious levels very quickly. Indeed, the fact that ratite eggs are incubated at 36.0–36.5°C reflects the fact that incubators cannot maintain optimal incubation temperatures towards the end of incubation when metabolic heat production is at its highest.

Further investigation of the effects of temperature during incubation in particular is crucial in understanding how the large eggs of ratites can be incubated efficiently. Whether any of the behavioural or health problems observed in ratite chicks during the first few days of their life are related to the incubation environment requires careful study.

4.5 Rearing

From here on I will consider those factors that impact on the welfare of ratite chicks during the first 3 months of life (refer Chaps. 1, 6 and 10). During this period, the birds are relatively vulnerable and it is known that mortality can be high during this period (Verwoerd et al. 1999). Much of the literature during the 1990s was concerned with how to farm ratites, rather than how to farm ratites to maximise their welfare. Unfortunately, welfare considerations (of any age of bird) have not been given the highest considerations in many parts of the world. Concerns over welfare (Wöhr and Erhard 2005) were often drawn from knowledge of other species on the assumption that these were directly applicable to ratites. When this was almost certainly a reasonable approach there did not seem to be sufficient impetus to investigate ratites directly with reference to their husbandry requirements to maximise welfare. Early considerations of welfare in Britain (Bertram 1993) were laudable but were not supported by targeted research that served to investigate how ratites were similar to, or different from, other farmed species. Concerns about keeping ostriches in Germany (Wöhr and Erhard 2005) were not based on scientific research in a European climate. However, as the interest in farming ratites waned during the late 1990s both in the USA and Europe (Deeming 2002b), the impetus to investigate the appropriate conditions was lost. For this reason, scientific investigations of the welfare of ratite chicks are rare (Mitchell 1999), but in subsequent sections the major welfare issues are discussed.

4.6 Behaviour

4.6.1 General Behaviour

Recording of behavioural activity under known conditions can inform the process of developing rearing systems that maximise welfare of the birds (refer Chaps. 6 and 10). To date descriptions of behaviour of young ratites are limited and almost exclusively carried out on ostriches.

Under natural conditions ostrich chicks are part of a clutch, and crèches develop when clutches combine (Bertram 1992). In southern Africa, commercial ostrich rearing often mimics this behaviour by fostering groups of chicks with adults in pens (Sales and Smith 1995; Huchzermeyer 1998). Data from Zimbabwe (Foggin and van Niekerk 1995) suggest that fostering is more productive than artificial rearing. Fostering in rheas is a useful alternative to artificial rearing conditions (Lábaque et al. 1999; Barri et al. 2005) but to the author's knowledge its value in emu farming has not been reported. Hence it can be deduced that social activity of ratite chicks is probably very important. The nature of this social structure has not been studied in any detail but it is known that ostrich chicks separated from their companions issue a warbling call (Deeming et al. 1996) that is rarely heard when the birds are in groups. Rhea chicks also exhibit a separation whistle together with other vocalisations that are extinguished by 7 weeks of age (Beaver 1978). Such vocalisations may be useful in assessment of welfare of young ratite chicks.

Time-budget analysis of young ostrich chicks showed that there are key behaviours that predominate in the activity of birds up to 14 days of age. Bubier et al. (1996) reported that locomotion occupied 23% of the time of chicks reared under artificial conditions with standing an additional 7%. Locomotion was a significant part of the activity (refer Chap. 8) of both naturally reared and artificially reared chicks (over 30% in both cases; Fig. 4.1) although Wöhr and Erhard (2005) did not record standing. Over a 14-day period walking was initially low but increased slowly in artificially reared chicks to over 40% of observations on day 6 but naturally reared chicks were more active earlier, reaching an observation rate for walking of over 40% on day 5 (Wöhr and Erhard 2005).

Resting is also a key component of early chick activity, with Bubier et al. (1996) reporting a rate of 11.2% of sitting under a heat lamp; resting was higher in the German study (Wöhr and Erhard 2005) with averages of 35–38% for chicks (Fig. 4.1) either resting (artificial rearing) or resting and underneath parents (naturally reared). In naturally reared chicks, resting under the parents was a high percentage of the time for the first 5 days but was almost absent thereafter. Lying down in artificially reared birds was also high for the first 3 days but declined thereafter, although it persisted in some chicks up to day 14.

Reports of behaviour in other ratite species are limited to that of emu chicks kept on an American farm (Elston et al. 1998). Behaviour was observed during the first and second weeks of life. Almost 50% of the time of the budget of chicks during week 1 was occupied by huddling, sleeping and walking, with feeding observed less

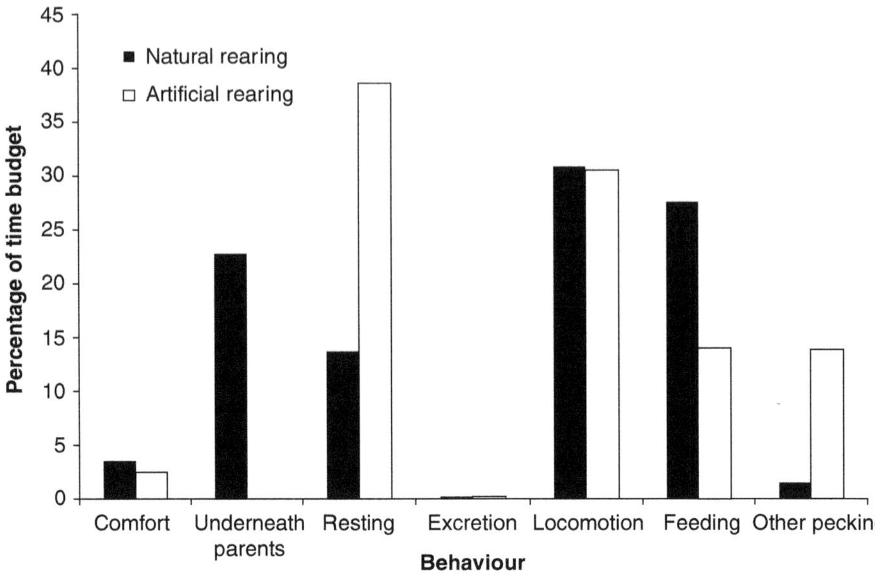

Fig. 4.1 Time budget of ostrich chicks reared with adult birds or under artificial conditions. Data from Wöhr and Erhard (2005)

than 10% of the time. During the second week, the amount of huddling was largely unchanged but walking and sleeping decreased to ~2% each. Feeding occupied almost 25% of the time budget and larger chicks had access to, and spent almost 9% of their time outside.

4.6.2 *Feeding Behaviour*

The interest in ratite farming during the 1990s led to a proliferation of ostriches and emus kept in captivity often by people inexperienced with keeping birds and advised by professionals unfamiliar with ratites. Such a situation led to some undesirable and incorrect advice. In particular, Kocan and Crawford (1994) recommended that, to promote utilisation of the yolk sac, food and water should be withheld from ostrich chicks until they were 6–8 days of age. The lack of access to water in particular poses a significant welfare issue. Also, providing food to day-old birds has been shown in poultry to promote yolk sac utilisation rather than retard it (Noy and Sklan 1997; Noy et al. 1996). Access to food and water from the earliest time should be priority even if they do not necessarily take advantage of them.

Feeding from the floor represented the biggest part (28%) of the time budget of chicks rearing in a closed barn whereas feeding from a food bowl (3.5%) was less common (Bubier et al. 1996). In a second study at the same site it was shown that pecking activity at food scattered on the ground increased as chick grew older (to 45 days of age) whereas pecking at the bowl declined (Paxton et al. 1997). Feeding from the environment in the German study ("grazing") was higher in naturally reared chicks (Fig. 4.1) and was shown to develop earlier and to a greater level over the first 14 days (Wöhr and Erhard 2005). Ostrich chicks exhibit a preference to peck at green objects (Lambert 1995) and when given the choice of lucerne (alfalfa) or pellets presented in an arena, young chicks (< 21 days) spent 74% of their time in the lucerne sector of the arena compared with 10% and 16% of the time in the control area without food or the area with pelleted food respectively (Deeming et al. 1996). Food selection appears to be well developed in even young birds. Cooper and Palmer (1994) showed that 21-day-old ostrich chicks exhibited significant preferences for particular plants and rejected two-thirds of the species offered.

Provision of food in the correct manner is crucial to ensure that intake is maximal. Deeming et al. (1996) suggest that scattering pelleted food and chopped vegetation on the floor provides the ideal stimulus for ostrich chicks to feed. Obtaining feed from a bowl is not a natural behaviour, although the birds will learn to associate food with particular food bowls. If the bowl is not transferred from one pen to another then this will induce misdirected pecking and can lead to gut impaction if excessive amounts of soil or bedding are consumed (Deeming et al. 1996). In general, many problems with rearing ostrich chicks are associated with poor rates of feeding and although this has not been investigated in emus or rheas, this is probably true for these species. These patterns of behaviour reflect what should be expected of ratites under normal rearing conditions when stressors are minimal. When other behaviours, such as misdirected pecking, gain prominence and impact on core activity, this may reflect an increased effect of stressors in the environment experienced by birds and hence indicate a reduction in welfare conditions.

4.6.3 Misdirected Pecking

In the German study (Wöhr and Erhard 2005), other forms of pecking, i.e. at objects other than food items, were very low in the naturally reared chicks but were 10 times greater in artificially reared birds (Fig. 4.1). Lambert et al. (1995) observed in ostrich chicks that pecking at things other than food was over 10% of the time of week-old birds and increased to over 12% in the second week, although the object of most pecking changed from the straw to the ground. Bubier et al. (1996) found that artificially reared chicks pecked at non-food items for around 10% of the time. For ostrich chicks between 26 and 33 days of age, pecking at non-food items was significantly negatively correlated with pecking at food items (Paxton et al. 1997). Paxton et al. (1997) reported that there were significant pen effects for pecking behaviour of ostrich chicks and suggested that there was "pen culture" associated with deleterious behaviours being spread between birds.

Feather pecking can start early on in life (Samson 1996) and was reported by Wöhr and Erhard (2005) at 0.03% in naturally reared chicks compared with 0.78% in artificially reared birds. Lambert et al. (1995) studied ostrich chicks up to 5 weeks

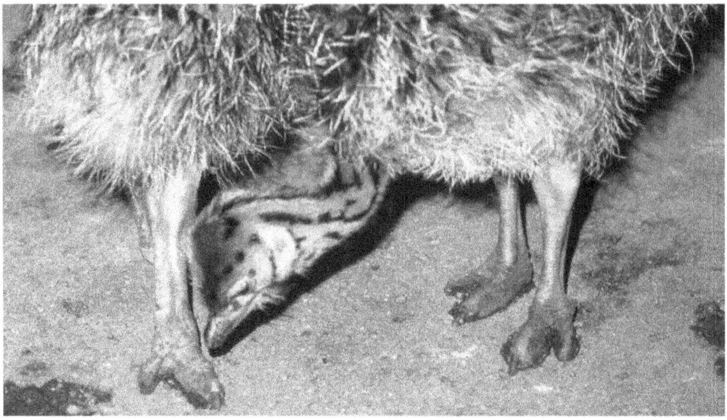

Fig. 4.2 Toe pecking in an ostrich chick. Photograph by D.C. Deeming. Reproduced with permission from Deeming et al. (1996) courtesy of Ratite Conference Books

of age and recorded the effects of pecking of individuals at other birds. Individual birds were observed to peck at toes (Fig. 4.2) and/or heads of companions and these were seen as problem behaviours associated with individuals (refer Chap. 6). Toe pecking was observed in all but two of 24 chicks and although initially seen at high rates at 10 days of age this was generally abolished by day 20. By contrast, head pecking was exhibited by only a few birds and reached a peak between 15 and 25 days of rearing. Pecking by these birds did not appear to represent attempts to establish a hierarchy within a group and the number of pecks delivered was negatively correlated with the growth rate of the birds. Paxton et al. (1997) also observed head pecking but this was associated in only 1–2 individuals. Although considered by Lambert et al. (1995) and Samson (1996) as aggressive, this was deemed inappropriate and this concept was rejected by Paxton et al. (1997). Rather, such pecking is better considered as misdirected feeding behaviour and may have better reflected the response of these birds to undefined stressors under less than optimal rearing conditions (Deeming and Bubier 1999). It was noted that once food was scattered on the floor of the pen such pecking declined (Paxton et al. 1997).

Pecking at faeces is a normal behaviour of ostrich chicks (Huchzermeyer 1998) and this is seen as a way of allowing for microbial colonisation of the gut (Huchzermeyer 1998; Samson 1996). However, many farmers and veterinarians saw this as undesirable given the possibility of bacterial or parasitic infections (Deeming et al. 1996). Coprophagy is impossible to prevent but can be minimised by regularly scattering pelleted food or chopped vegetation on the floor of the pen. Use of a probiotic, such as natural yoghurt, has been used in an attempt to colonise the gut of chicks with beneficial species of bacteria (Deeming et al. 1996) although the efficacy of this was not investigated. A commercial probiotic was shown to have beneficial effects on growth and feed conversion of ostrich chicks grown up for 9 weeks (Foroudi et al. 2008), but no significant benefits of probiotics have been demonstrated in rhea chicks (Gri et al. 2008).

4.7 Environmental Conditions and Management

Many ratite chicks are reared under artificial conditions and are exposed to stimuli that are novel. In particular, human activity is often very high in such conditions involving routine feeding and cleaning. Elements of these conditions may act as stressors to birds and so affect their welfare. Paxton et al. (1997) studied whether barrier height affected ostrich chick behaviour on the premise that high barriers (60 cm high versus 30 cm) would limit the visual field of the birds and so reduce exposure to human activity. Pecking activity towards food, either in bowls or scattered on the floor, or directed at other objects in the pen was not significantly affected by the height of the barriers.

Positive attempts to stimulate ratite chicks in artificial rearing environments can be seen as improving activity of birds and perhaps increasing the welfare of individuals. Pet toys and white plastic spoons were seen by Deeming et al. (1996) as being a useful method of stimulating and increasing activity of ostrich chicks but no formal studies of the effects were reported. Scattering chopped vegetation on a food bowl (Cooper 2000) or more widely around a pen has the added advantage of improving nutrition as well as stimulating activity (Deeming et al. 1996; Paxton et al. 1997).

The effects of environmental enrichment with chopped cabbage were investigated by Christensen and Nielsen (2004). In this study chicks were reared from hatch in chipboard-floored pens left barren (control) or provided with chopped cabbage, coniferous cones and willow twigs. There was also access to an outside pen. Pecking activity (pecks per chick per min) was recorded from day 10 to day 21 post-hatching. The mean number of pecks per chick did not differ between the groups but enriched chicks directed 22% of all pecks towards the cabbage and largely ignored the cones and sticks. Almost 60% of the pecks by control birds were directed towards the chipboard floors compared with only 31% for the enriched birds. Pecking rates at other objects were not significantly different, although 50% of the control chicks spent 0.4–12% of the time pecking at a plastic door at the entrance to the outdoor part of the pen; such behaviour was not observed in enriched chicks. Feed consumption was significantly ($P < 0.01$) higher for the enriched chicks in addition to consumption of 23–29 g of cabbage per chick per day. The effects of this on growth rate of the birds were not recorded.

Christensen and Nielsen (2004) also tested the reaction of ostrich chicks to novel objects placed in the pen. When confronted with stems of sorrel (*Rumex*), enriched chicks spent 27% of the first 30 min within 0.25 m of the object compared with 0% for the control birds. Pecking at the object was also higher in the enriched chicks. By contrast, there was no significant difference between the groups in their response to adult ostrich feathers. It was concluded (Christensen and Nielsen 2004) that enrichment improves the welfare of ostrich chicks by increasing exploration and reducing pecking directed at fixtures in the pen without compromising food consumption.

In wild situations, clutches of ostrich chicks combine to form crèches that can be large in size yet be accompanied only one pair of adults (Bertram 1992; Deeming and Bubier 1999). Under commercial rearing chicks are often mixed together or sorted by size (Deeming and Ayres 1994). Kamau et al. (2002) investigated whether the mixing of flocks of chicks created any problems in chick rearing. After translocation and mixing of chicks the heterophil-to-lymphocyte ratio increased, which was interpreted as indicating that this process was stressful. It is unclear whether it was the mixing of birds or the translocation to another pen that was the stressor (Kamau et al. 2002) and further work is needed to clarify this point.

4.8 Temperature Relations

At hatch, ratites chicks are large (~700–900 g for ostriches, ~400 g for emus and rheas) compared with poultry species and can easily be considered as well able to cope with relatively cool temperatures during rearing. Indeed Brown and Prior (1999) demonstrated that day-old ostrich chicks can maintain a body temperature over 36°C at ambient temperatures of 20°C. Although it was suggested that such advanced homeothermy precluded a need to rear chicks at high temperatures (30°C), Brown and Prior (1999) did recognise that birds could develop hypo- or hyperthermia, and that thermoregulatory behaviour is an important means by which chicks can reduce energy expenditure.

Seeking external sources of heat is a key component of the time budget of naturally reared and artificially reared ostrich chicks, which seek the brooding adult (Wöhr and Erhard 2005) and heat lamps (Wöhr and Erhard 2005; Bubier et al. 1996), respectively. During the first 2 days post-hatching ostrich chicks allowed to stray from the immediate vicinity of a heat lamp get chilled and exhibit high mortality (Deeming et al. 1996). Thereafter, if the room temperature is too cold ostrich chicks will spend a lot of time under the heat lamp rather than walking and feeding. This has implications for weight gain and normal development of the leg muscles. Conversely, if the temperature conditions are too warm ostrich chicks can thermoregulate by exposing bare skin under the wings (Fig. 4.3). In general, failure to provide a comfortable thermal environment means that the chicks spend more time seeking heat and less time feeding and this could reduce their welfare. A better understanding of the temperature requirements of the birds comes from directly observing the birds and responding accordingly.

4.9 Handling

Unless care is taken ratite chicks are vulnerable to stress or physical injury during handling. The general advice for young birds is for handlers to be gentle and considerate (refer Chap. 1) and properly trained (Mitchell 1999) (refer Chap. 6). Deeming et al. (1996a) gave recommendations for handling chicks. Young chicks can be picked up by placing the hand under the body with the legs free to move

Fig. 4.3 Thermoregulation in a chick at 7 days of age. Note that the wings are held forward to expose bare skin on the flanks. Photograph by D.C. Deeming. Reproduced with permission from Deeming et al. (1996) courtesy of Ratite Conference Books

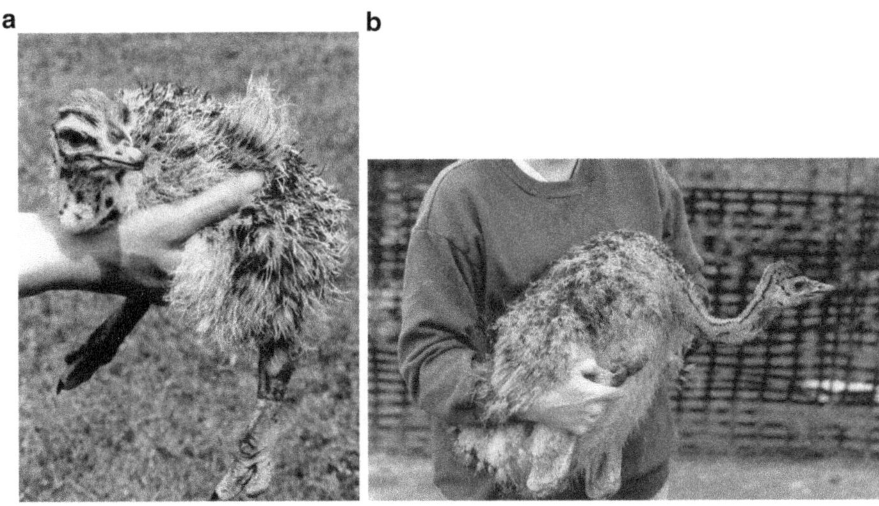

Fig. 4.4 How to handle an ostrich chick at day-old (a) and 5 weeks of age (b). Photographs by D.C. Deeming. Reproduced with permission from Deeming et al. (1996) courtesy of Ratite Conference Books

(Fig. 4.4a). When putting the bird down it should be able to find its feet before being released, thereby preventing injury. Regular handling from an early age habituates the birds to people and so making handling later less stressful. Older chicks can be picked up by restraining the legs and the biggest chicks can be retained by standing astride them (Fig. 4.4b). In general, avoiding sudden movements will minimise chicks becoming frightened and running off, which is a common cause of slipping on a wet floor or on faeces, potentially causing damage to the leg bones (refer Chaps. 4 and 9). These methods are based on observations under farming conditions but there has been no scientific study of how inappropriate handling can affect chick welfare.

Artificial rearing requires considerable visual contact with humans but little is known whether this constitutes a stressor. Wearing the same clothes is seen as minimising stress in ostriches during rearing (Huchzermeyer 1998) although no physiological measurements have been reported to support this idea. The response of ostrich chicks to people has only been reported once (Deeming and Bubier 1999). From hatching through to 2 and 4 months of age ostrich chicks increased standing behaviour and decreased feeding activity but were otherwise neutral to a person next to the pen. In chicks at 6 months or older, interest in humans increased so that they spent significantly more of their time in the half of the enclosure closest to the person.

4.10 Transportation

A major issue with regard to chick welfare is the transport of animals between farms (refer Chaps. 10 and 11). Wotton and Hewitt (1999) reviewed ostrich transport in adults and chicks, although for the latter the comments were generalised and aimed towards best practice. Hence, there are descriptions of how to handle and load the birds, why bedding is important, control of temperature and ventilation and journey time. The authors had little information to draw upon and were reliant on practical guides or advice from farmers. Since the turn of the twenty-first century, there has been little improvement in our understanding of transport of ostrich and other ratite chicks.

Although considered to be of importance (Mitchell et al. 1996), the physiological effects of transport on ostrich chicks are poorly reported. The effects of transporting ostrich chicks aged around 25 days in chick boxes for 950 km in July in Italy was investigated by Piccione et al. (2005). Internal lorry temperature increased from 28°C to a maximum of 33°C over the 12-h trip and the chicks exhibited a rise in cloacal temperature of over 0.5°C upon arrival and a 7% drop in body mass compared with initial measurements pre-transport. Restoration of these parameters to pre-transport values took at least 4 days post-transport. Almost 17% of the chicks in the study died during transport. These results were interpreted by Piccione et al. (2005) as indicating that the process was a stressor and its effects persisted in the short term after the stressor was removed.

A study of the transport of 3-month-old ostrich chicks in New Zealand showed that whether the birds were sitting or standing was crucial in determining heart rates and skin temperatures (Crowther et al. 2003). These birds were housed in individual pens in a horse trailer and transported for 7–8 h during the day and night. Birds that sat during transport had lower heart rates and skin temperatures than when they were standing. The latter condition may have in part been reflected by the cooler temperatures experienced by the birds at night, when sitting was most prevalent and provided the greatest stability for the birds. Crowther et al. (2003) did not record the recovery times of birds but did recommend that transport in the dark was the preferred option. In Nigeria, transportation of slightly older ostrich chicks

(4-months) over 224 km that took approximately 4 h had more transient physiological effects (Minka and Ayo 2007). During transport, rectal temperature, respiratory rate and heart rate all increased but returned to pre-transport levels within 6 h of unloading. The process was in a darkened vehicle and during the early hours of the day and this may have mitigated against the stress of transport.

4.11 Health Issues

The health of ratite chicks has posed serious problems in development of farming and this has implications for their welfare. Aside from proper rearing conditions (see above), the health of the flock is of critical importance (refer Chap. 9). Space precludes a comprehensive description of the health problems of ratite chicks but several good reviews are available (Verwoerd et al. 1999; Huchzermeyer 1998, 1999; Speer 1996; Perelman 1999). Ostrich chicks clearly exhibit behavioural characteristics when they are ill (see Fig. 4.5).

4.11.1 Growth

Variation in the growth rate of ostriches (Fig. 4.6) is commonly reported (e.g. Deeming and Ayres 1994; Verwoerd et al. 1999; Deeming et al. 1993; Dawson et al. 1996) and appears to be typical of birds reared under farming conditions. In South Africa, chicks are routinely sorted by weight at 3 months of age (Meyer et al. 2003). Data are available for the growth rates of emus (Goonewardene et al. 2003) and rheas (Barri et al. 2005) but there is little comment on the variation observed. It is

Fig. 4.5 Typical appearance of a "sick" chick. Illustration by Richard Tibbitts. Reproduced with permission from Deeming et al. (1996) courtesy of Ratite Conference Books

Fig. 4.6 Example of the extremes in ostrich chick size. Both birds are 6 weeks of age and reared under the same conditions [as described by Deeming et al. (1993)]. Photograph by D.C. Deeming

suspected that the high level of variation observed in ostriches is also found in these two species. Factors that influence rates of growth are varied but include rates of feeding (see above), abnormal pecking activity (see above) and health problems (see below).

Although there appear to be many factors that influence the rate at which ratite chicks grow, one important factor is the low level of endogenous thyroxine. Ratites are generally considered to exhibit neoteny, which would correspond with the low levels of plasma of thyroxine, which are atypical of birds (Dawson et al. 1996). Compared to other species of bird, thyroid hormones are naturally very low in juvenile ostriches (Dawson et al. 1996) and emus (Blache et al. 2001). Low body weight cannot be directly attributable to low levels of thyroxine in ostriches or emus. The correlation reported by Dawson et al. (1996) was the result of the order in which chicks were caught and blood sampled (Dawson and Deeming 1997) – big chicks were caught first and so were confined for less time than the smaller birds. Imposition of such stress on birds rapidly lowers the thyroxine level and the effect is time-dependent. Indeed, in ostriches the initial levels are so low in the first instance that even imposition of a short-term stressor has a very dramatic impact of lowering thyroxine levels to almost zero (Dawson and Deeming 1997). Therefore, imposing a stressor on ratite chicks can have a significant and prolonged physiological effect that is much exaggerated compared to that seen in other bird species.

Adoption of chicks by adult ostriches living in paddocks is commonly observed in South Africa (Verwoerd et al. 1999) although the effects on mortality or growth rates have not been documented. One attempt to improve rates of growth in rhea chicks used an adoption system, so that adult males rear the chicks under semi-natural conditions. Although time to full adoption was positively correlated with the initial brood size cared for by the male, mortality of chicks was comparable under adoption relative to those seen under typical intensive rearing systems (Lábaque et al. 1999; Barri et al. 2005). There were significant benefits in chick growth rates over the first 3 months of life by adoption during the early part of the breeding season (Barri et al. 2005). Later in the season growth rates of adopted chicks were not significantly higher.

4.11.2 Yolk Sac Problems

Yolk sac infection is often problematic in ratites (Huchzermeyer 1998; Speer 1996). Chicks fail to thrive and die in an emaciated state within 2–3 weeks of hatching (refer Chap. 9) but with a large yolk sac (Deeming 1995b). Diagnosis is typically that there is failure of the bird to utilise the yolk and that it is this that kills the chick. A common solution to this problem in the 1990s was surgical removal of the yolk sac, although the prognosis for treated chicks was very poor. Starved poultry chicks utilise their residual yolk slower than fed chicks (Noy and Sklan 1997; Noy et al. 1996). Deeming (1995b) argued that the retained yolk sac was not the cause of the problem but rather a symptom of another problem, which was typically not being investigated because of the diagnosis of yolk sac infection. Deeming (1995b) showed that the rates of yolk utilisation in sterile and contaminated yolk sacs were the same. Indeed in first grade, healthy, day-old poultry and waterfowl chicks considerable numbers of bacteria can be isolated from the yolk sac without any indication of disease in the rest of the flocks (Deeming 1995).

4.11.3 Leg Problems

Ratite chicks are long-legged species and the health of the legs is critical for the individual birds (refer Chap. 9). There are a variety of common leg deformities in ratites (Verwoerd et al. 1999; Huchzermeyer 1998, 1999; Speer 1996; Dick and Deeming 1996; Mushi et al. 1999), all of which have adverse impact on bird health and welfare. These include splayed legs, tibiotarsal rotation, rolled toes and slipped tendons (Fig. 4.7). The problems caused vary in their effects. Tibiotarsal rotation and slipped tendons produce such severe physical damage and movement problems that they are invariably fatal because the bird stops eating and loses condition, or is culled. Splayed legs can be relatively easy to resolve but rolled toes produce the most long-lasting problem. In this case, the pad that underlies the large toe in ostriches comes to lie on the median side so that the bird is walking on skin overlying bone (Fig. 4.7b). Deeming et al. (1996) reported that exercise and good consumption of feed could resolve this problem, although in the most severe cases the phalangeal bones are twisted and the deformity is permanent.

Despite much speculation, the cause of each of these leg problems has not really been elucidated. For tibiotarsal rotation physical damage to the cartilaginous growth

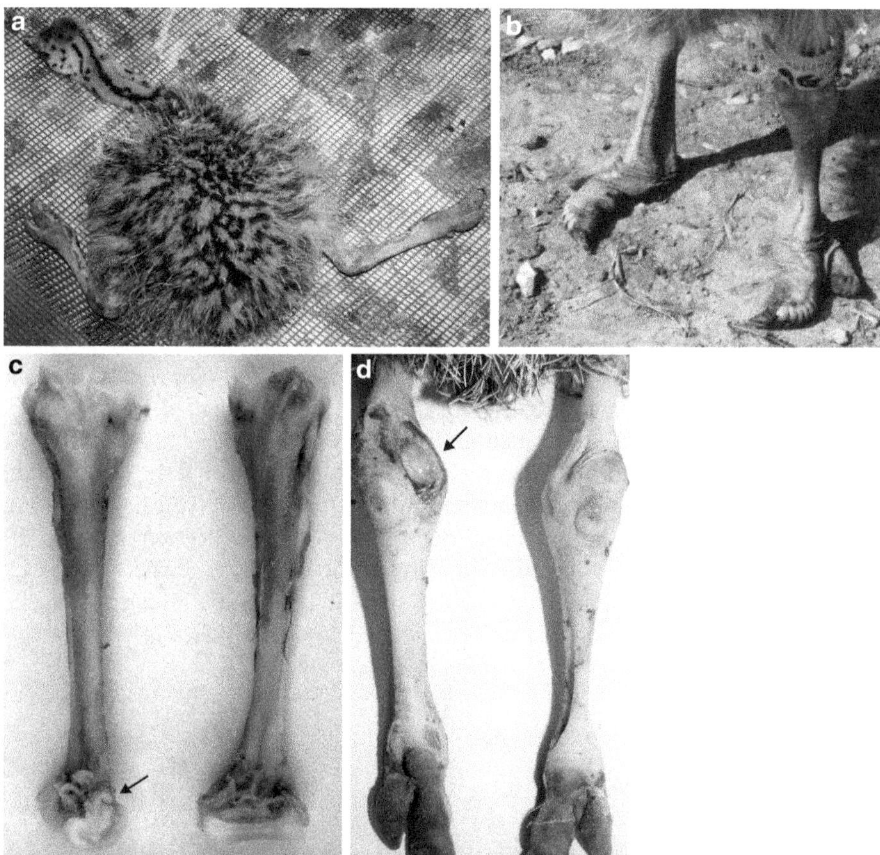

Fig. 4.7 Examples of the pathology of leg problems in ostrich chicks. (**a**) Day-old chick exhibiting splay legs; (**b**) Bilateral rolled toes; (**c**) Tibiotarsi from a chick exhibiting rotation of the right leg – note the distal face of the bone is turned outwards by 70° (*arrow*); (**d**) Compound dislocation of gastrocnemius ("slipped") tendon (*arrow*). Photographs by D. C. Deeming. Reproduced with permission from Deeming et al. (1996) courtesy of Ratite Conference Books

plates of the distal tibiotarsus has been suggested as being the cause (Deeming et al. 1996), but no real understanding of the other problems exists. This is an issue for ratite welfare given the high incidence of the problems and the need for focussed research to determine the causes of these problems remains. If a diagnosis could be reached for any of the problems development of suitable preventative measures or treatments would lead to a significant improvement in welfare of ratite chicks.

4.12 Declawing

The skin of ostriches and emus is commercially important (refer Chap. 6) and its value can be significantly decreased by the presence of scratches and other marks (Meyer et al. 2002). As a result surgical amputation of the phalangeal joint ("declawing") is regularly performed on ostriches and emus in South Africa and Australia as a commercial management tool (Mushi et al. 1999; Glatz 2005). Declawing may reduce downgrading of skins at slaughter (Meyer et al. 2002; Glatz 2005) but it may affect chick welfare on the day of treatment and potentially causes chronic pain and pose other health issues (Glatz 2005).

In Australia Glatz (2005) recorded the behaviour of 1-day-old ostrich chicks for 12 h (overnight) after declawing and compared it to untreated controls but it is unclear whether the brooder room was well lit or just relied on the heat lamp for illumination. There were no significant effects of declawing on the number of bouts of eating, drinking and preening. However, there were significant reductions in the numbers of bouts of sitting and standing in declawed birds. Additionally, bouts of pecking at the environment and potentially agonistic behaviours to companions were significantly reduced in the declawed birds. By 1 week of age, the diurnal behaviour of control and declawed chicks showed no significant difference except for higher levels of sitting and standing in the outside run (Glatz 2005). By 9 months of age, the time budget of declawed ostriches in a South African study was not significantly different to that of control birds except for a higher incidence of pecking at objects in the former group (Glatz 2005). Declawing had no effect on the growth rate of chicks from hatch to slaughter (Glatz 2005).

Despite the conclusions that declawing seems to be unproblematic (Meyer et al. 2002; Glatz 2005) there are concerns. Glatz (2005) defined a "bout" as a chick performing any particular behaviour for 5 s or more but, despite having video of the chicks, actual time budgets were not recorded. As a result Glatz (2005) could only speculate whether periods of sitting were longer for declawed 1-day-old chicks. It is possible that the actual periods of sitting and standing by declawed ostriches were longer and shorter, respectively, compared to the controls. Whilst declawing does have commercial benefits evidence that Glatz (2005) presents to suggest that the procedure does not have any immediate welfare problems needs re-visiting to clarify the actual effect on the time budget of chicks immediately after declawing. The effect of declawing on ratite chick welfare needs further investigation.

4.13 Conclusion

In conclusion, it is unclear whether welfare considerations can directly apply to ratite embryos. Adoption of commercial practises that maximise hatchability and chick quality can only serve to optimise the environment for the development of embryos and so not compromise their welfare. The ability of an embryo to perceive its environment *in ovo* is not fully understood and possible mechanisms by which it could elicit a change in the incubation conditions have not been demonstrated. It is only after internal pipping and thereafter during which the bird is hatching, that we can demonstrate that through vocalisations an embryo can communicate with its parent and so change its behaviour. It is from this point that welfare considerations of embryos and chicks certainly need to be considered.

Welfare of ratite chicks has received little direct attention or scientific research but this should not detract from its importance. To date research has been primarily limited to ostriches but further work is needed for all three commercially important species to ensure that species differences can be accommodated. Therefore, whilst there are practices, such as declawing, that are peculiar to ratites, in general welfare considerations that apply to poultry or waterfowl species should be largely applicable to these larger species. There is a clear need for more scientific research into the factors that may impact either negatively or positively on ratite chick welfare. In particular, there is a need to better understand rearing conditions and how they impact upon growth rates of chicks. The physiological effects of possible stressors would also be a useful area for research. Whilst declawing appears to have little long-term impact on chick welfare, there is an urgent need to better understand the physiological and physical effects of such mutilation performed for commercial gain.

One problem with such research is that sources of financial support may not be forthcoming from organisations other than the commercial operations themselves. In other livestock farming, it is known that welfare concerns have an impact on the management and profitability of businesses. Ratite farming cannot be immune from this and so there is a need for an increased emphasis on good welfare within the ratite industry. Trade associations would be well served to fund studies into chick welfare because in the long term our increased understanding of how to rear these birds under appropriate conditions should have positive economic impacts through increased productivity and profitability.

References

Barri FR, Navarro JL, Maceira NO, Martella MB (2005) Rearing Greater Rhea (Rhea americana) chicks: is adoption more effective than the artificial intensive system? Br Poult Sci 46:22–25

Beaver PW (1978) Ontogeny of vocalization in the greater rhea. Auk 95:382–388

Bertram BCR (1992) The ostrich communal nesting system. Princeton University Press, Princeton, USA

Bertram B (1993) Welfare standards for the humane farming of ostriches in the United Kingdom. RSPCA, Horsham, UK
Blache D, Blackberry MA, Van Cleeff J, Martin GB (2001) Plasma thyroid hormones and growth hormone in embryonic and growing emus (Dromaius novaehollandiae). Rep Fertil Dev 13: 125–132
Boerjan M (2002) Programs for single stafe incubation and chick quality. Avian Poult Biol Rev 13:237–238
Brown CR, Prior SA (1999) Development of body temperature regulation in ostrich chicks. Br Poult Sci 40:529–535
Brua RB (2002) Parent-embryo interactions. In: Deeming DC (ed) Avian incubation: behaviour, environment and evolution. Oxford University Press, Oxford, UK, pp 88–99
Bruning DF (1973) Breeding and rearing rheas in captivity. Int Zool Yearb 13:163–174
Bubier NE, Lambert MS, Deeming DC, Ayres LL, Sibly RM (1996) Time budget and colour preferences (with special reference to feeding) of ostrich (Struthio camelus) chicks in captivity. Br Poult Sci 37:547–551
Christensen JW, Nielsen BL (2004) Environmental enrichment for ostrich, Struthio camelus, chicks. Anim Welf 13:119–124
Cooper RS (2000) Management of ostrich (Struthio camelus) chicks. World Poult Sci J 56:33–44
Cooper SM, Palmer T (1994) Observations on the dietary choice of free-ranging juvenile ostriches. Ost 66:251–255
Crowther C, Davies R, Glass W (2003) The effects of night transportation on the heart rate and skin temperature of ostriches during real transportation. Meat Sci 64:365–370
Dawson D, Deeming DC (1997) Thyroid function in ostriches compared to that in non-ratite birds. In: XIII international congress of comparative endocrinology, Yokohama, Japan, 16–21 Nov 1997, pp 439–444
Dawson A, Deeming DC, Dick ACK, Sharp PJ (1996) Plasma thyroxine levels in farmed ostriches (Struthio camelus) in relation to age, body weight and growth hormone. Gen Comp Endocrinol 103:308–315
Deeming DC (1993) The incubation requirements of ostrich (Struthio camelus) eggs and embryos. In: Proceedings Meeting Australian Ostrich Association, Proceeding 217 Post Graduate Comm Veterinary Science, University of Sydney, Sydney, Australia, pp 1–66
Deeming DC (1995a) The hatching sequence of ostrich (Struthio camelus) embryos with notes on development as observed by candling. Br Poult Sci 36:67–78
Deeming DC (1995b) Possible effects of microbial infection on yolk utilisation in ostrich chicks. Vet Rec 136:270–271
Deeming DC (1997) Ratite egg incubation: a practical guide. Ratite Conference Books, High Wycombe
Deeming DC (2002a) Avian incubation: behaviour, environment and evolution. Oxford University Press, Oxford
Deeming DC (2002b) The fall and rise of ratite production. In: Proceedings of 11th European Poultry Conference, Bremen, Germany, 6–10 Sept 2002, CD-ROM
Deeming DC (2005) Yolk sac, body dimensions, and hatchling quality of ducklings, chicks and poults. Br Poult Sci 46:560–564
Deeming DC, Ar A (1999) Factors affecting the success of commercial incubation. In: Deeming DC (ed) The ostrich: biology, production and health. CAB International, Wallingford, pp 159–190
Deeming DC, Ayres L (1994) Factors affecting the rate of growth of ostrich (Struthio camelus) chicks in captivity. Vet Rec 135:617–622
Deeming DC, Bubier NE (1999) Behaviour in natural and captive environments. In: Deeming DC (ed) The ostrich: biology, production and health. CAB International, Wallingford, pp 83–104
Deeming DC, Ayres L, Ayres FJ (1993) Observations on the first commercial production of ostrich (Struthio camelus) eggs in the UK: rearing of chicks. Vet Rec 132:627–631
Deeming DC, Dick ACK, Ayres LL (1996) Ostrich chick rearing. A stockman's guide. Ratite Conference Books, Oxfordshire

Dick ACK, Deeming DC (1996) Veterinary problems encountered on an ostrich farms in Great Britain. In: Deeming DC (ed) Improving our understanding of ratites in a farming environment. Ratite Conference, Oxfordshire, pp 40–41
Elston JJ, Beck MM, Scheideler SE (1998) Behavioural analysis of emu chicks and breeding adults. In: Ratites in a competitive world. Proceedings of 2nd international ratite congress, Oudtshoorn, South Africa, 21–25 Sept 1998, pp 177–180
Foggin C, Van Niekerk A (1995) Ostriches in the wild, colony breeding and foster rearing. In: Proceedings of 5th Australian Ostrich Association Conference, Goldcoast, Australia, pp 111–116
Foroudi F, Asfar A, Kolli ME (2008) The effect of dietary probiotic supplementation on production performance of ostrich chicks. Aust J Exp Agric 48(10):12–13
Gefen E, Ar A (2001) Gas exchange and energy metabolism of the ostrich (Struthio camelus) embryo. Comp Biol Phys A130:689–699
Glatz PC (2005) Effect of declawing on behaviour and skin quality of ostriches. In: Proceedings of the 3rd international ratite science symposium (WPSA), 14–16 Oct 2005, Madrid, pp 157–162
Goonewardene LA, Wang Z, Okine E, Zuidhof MJ, Dunk E, Onderka D (2003) Comparative growth characteristics of Emus (Dromaius novaehollandiae). J Appl Poult Res 12:27–31
Gri LR, Martella MB, Navarro JL (2008) Efficacy of probiotics in intensively reared Greater Rhea (Rhea americana) chicks. Arch Geflüg 72:136–139
Hill D (2001) Chick length uniformity profiles as a field measurement of chick quality? Avian Poult Biol Rev 12:188
Huchzermeyer FW (1998) Diseases of ostriches and other ratites. Agricultural Research Council, Onderstepoort Veterinary Institute, South Africa
Huchzermeyer FW (1999) Veterinary problems. In: Deeming DC (ed) The ostrich: biology, production and health. CAB International, Wallingford, pp 293–320
Kamau JM, Patrick BT, Mushi EZ (2002) The effect of mixing and translocating juvenile ostriches (Struthio camelus) in Botswana on the heterophil to lymphocyte ratio. Trop Anim Heal Prod 34:249–256
Kocan AA, Crawford JA (1994) An ostrich farmers handbook. The Ostrich News, Cache, Oklahoma
Lábaque MC, Navarro JL, Martella MB (1999) A note on chick adoption: a complementary strategy for rearing rheas. Appl Anim Behav Sci 63:165–170
Lambert MS, Deeming DC, Sibly RM, Ayres LL (1995) The relationship between pecking behaviour and growth rate of ostrich (Struthio camelus) chicks in captivity. Appl Anim Behav Sci 46:93–101
Lourens A, van den Brand H, Meijerhof R, Kemp B (2005) Effect of eggshell temperature during incubation on embryo development, hatchability and post-hatch development. Poult Sci 84:914–920
Meyer A, Cloete SWP, Brown CR, van Schalkwyk SJ (2002) Declawing ostrich (Struthio camelus domesticus) chicks to minimise skin damage during rearing. S Afr J Anim Sci 32:192–200
Meyer A, Cloete SWP, Brown CR (2003) The influence of separate-sex rearing on ostrich behaviour and skin damage. S Afr J Anim Sci 33:95–104
Minka NS, Ayo J (2007) Road transportation effect on rectal temperature, respiration and heart rates of ostrich (Struthio camelus) chicks. Veterinarski Ark 77:39–46
Mitchell MA (1999) Welfare. In: Deeming DC (ed) The ostrich: biology, production and health. CAB International, Wallingford, pp 217–230
Mitchell MA, Kettlewell PJ, Sandercock DA, Maxwell MH, Spackman D (1996) Physiological stress in ostriches during road transport. In: Deeming DC (ed) Improving our understanding of ratites in a farming environment. Ratite Conference, Oxfordshire, pp 79–80
Mushi EZ, Binta MG, Chabo RG, Isa JF, Phuti MS (1999) Limb deformities of farmed ostrich (Struthio camelus) chicks in Botswana. Trop Anim Heal Prod 31:397–404
Noy Y, Sklan D (1997) Posthatch development in poultry. J Appl Poult Sci 1:344–354
Noy Y, Uni Z, Sklan D (1996) Routes of yolk utilisation in the newly-hatched chick. Br Poult Sci 37:987–996

Paxton CM, Bubier NE, Deeming DC (1997) Feeding and pecking behaviour in ostrich (Struthio camelus) chicks in captivity. Br Poult Sci 38:151–155

Perelman B (1999) Veterinary problems. In: Deeming DC (ed) The ostrich: biology, production and health. CAB International, Wallingford, pp 321–346

Piccione G, Costa A, Giudice E, Caola G (2005) Preliminary investigation into thermal stress during diurnal road transportation of young ostriches (Struthio camelus). Arch Tierz Dummerstorf 48:194–200

Rogers LJ (1995) The development of brain and behaviour in the chicken. CAB International, Wallingford, p 22

Sales J, Smith WA (1995) Incubation and management. In: Smith WA (ed) Practical guide for ostrich management and ostrich products. Alltech, Stellenbosch University Printers, South Africa, pp 3–7

Samson J (1996) Behavioral problems of farmed ostriches in Canada. Can Vet J 37:412–414

Smit DJvZ (1963) Ostrich farming in the Little Karoo. Bulletin No. 358, Department of Agricultural Technical Services, Pretoria, South Africa

Speer BL (1996) Developmental problems in young ratites. In: Tully TN, Shane SM (eds) Ratite management, medicine and surgery. Krieger, Malabar, FL, pp 147–154

Tona K, Onagbesan O, Jego Y, Kamers B, Decuypere E, Bruggemann V (2004) Comparison of physiological parameters during incubation, chick quality and growth performance of broilers from three lines of broiler breeders differing in genetic composition and growth rate. Poult Sci 83:507–513

Tona K, Onagbesan O, De Ketelaere B, Bruggemann V, Decuypere E (2005) Interrelationships between chick quality parameters and the effect of individual parameter on broiler relative growth to 7 days of age. Av Poult Biol Rev 16:71–72

van Schalkwyk SJ, Cloete SWP, Brown CR, Brand Z (2000) Hatching success of ostrich eggs in relation to setting, turning and angle of rotation. Br Poult Sci 41:46–52

Verwoerd D, Deeming DC, Angel CR, Perelman B (1999) Rearing environments around the world. In: Deeming DC (ed) The ostrich: biology, production and health. CAB International, Wallingford, pp 191–216

Wöhr AC, Erhard M (2005) Ostrich farming in Germany – An animal welfare issue? In: Proceedings of the 3rd International Ratite Science Symposium (WPSA), Madrid, Spain, pp 145–156

Wotton SB, Hewitt L (1999) Transportation of ostriches – a review. Vet Rec 145:725–731

Chapter 5
Ostrich Nutrition and Welfare

T. Brand and A. Olivier

Abstract The digestive system of the ostrich is adapted to digest fibre-rich plant materials, which distinguishes it from other monogastric herbivores. Diets and feeding systems of ostriches should therefore be formulated to take this into account and to use this advantage to produce ostrich products in an economic way. The chapter summarises the nutrient requirements for growing and breeding ostriches in terms of metabolisable energy, protein and amino acid contents of the diets to optimise production. Attention is given to the feeding behaviour of ostrich chicks, which is important to ensure acceptable feed intake and normal feeding behaviour. Due to the unique digestive system of ostriches, it may also be kept on pasture which will have a definite economic advantage in the production system. Different feeding and welfare implications such as feed and water management, stockmanship, feeding space, feeding times and feeding conditions are discussed. Discussions on nutrition-related diseases include deficiencies and imbalances, clostridial overgrowth and mycotoxins.

Keywords Anatomy · Breeding birds · Chickens · Digestive system · Diseases · Feed intake · Feed processing · Feeding behaviour · Growing birds · Metabolisable energy · Nutrient requirements · Nutrition · Optimisation model · Pasture · Supplementation · Welfare

T. Brand (✉)
Department of Agriculture Western Cape, Elsenburg Animal Production Institute, Private Bag X1, Elsenburg, 7607 Stellenbosch, South Africa
and
Department of Animal Sciences, University of Stellenbosch, Stellenbosch 7600, South Africa
e-mail: TersB@elsenburg.com

A. Olivier
Ostrivet, Kooperasie Straat, Oudtshoorn 6620, South Africa
e-mail: aolivier@k3sa.com

P.C. Glatz et al. (eds.), *The Welfare of Farmed Ratites*, Animal Welfare 11,
DOI 10.1007/978-3-642-19297-5_5,

5.1 Introduction

The ostrich industry is small compared to animal industries such as the poultry, red meat and dairy industries. Nevertheless, the ostrich industry has a gross turnover of approximately R1.5 billion in South Africa, which provides 60–70% of ostrich products produced globally (Brand and Cloete 2007). Other countries which currently produce significant number of ostriches include Zimbabwe, Israel and Brazil. It is estimated that in South Africa approximately 23,000 female breeding birds are currently kept, which produce approximately 200,000 slaughter birds per year (South African Ostrich Business Chamber (SAOBC)). The main products of ostriches are skins, meat and feathers, and the ostrich industry benefits all participants in these associated industries.

The nutrition of ostriches is important as feed normally comprises approximately 75% of the total costs of an intensive ostrich production unit (Jordaan et al. 2008). The knowledge of ostrich nutrition is, however, still in its infancy compared to that of broiler and pig nutrition (Gous and Brand 2008). Simulation models that have been applied successfully to simulate food intake and growth of broilers and pigs (EFG Software 2008) are now also being applied to the subject of ostrich nutrition, and may help to accelerate progress. The current chapter is an attempt to summarise current and relevant knowledge on ostrich nutrition and nutritional diseases, which has relevance to the welfare of ostriches in modern ostrich production systems.

5.2 Anatomical Structures and Digestion

Ostriches are monogastric herbivores with a relatively large digestive tract, which enables them to utilise fibrous plant material. Their digestive tract consists of a beak and mouth, oesophagus, proventriculus (glandular stomach), gizzard (muscular stomach), small intestine (duodenum, jejunum and ileum), large intestine (two caeca and the proximal, middle and distal colon) and cloaca (Brand and Gous 2006a) (Fig. 5.1).

The ability of ostriches to digest fibre distinguishes it from other monogastric herbivores, with Swart (2008) clearly demonstrating the ability of ostriches to digest hemi-cellulose and cellulose. This ability, however, develops only after 10 weeks of age, while younger chicks do not digest fibre well (Nizza and Di Meo 2000; Salih et al. 1998; Angel 1996). The gizzard plays an important role in digestion, and stones and pebbles, 50–70% of the size of the toenail of the bird, should always be available to be ingested by the bird to help to mechanically grind the feed to a fine form. The development of the gastrointestinal tract of the ostrich switches dramatically from that of a typical bird neonate to that of a hindgut fermenter in the space of 70–80 days (Van der Walt et al. 2008) and diets may be changed accordingly. Ostriches, because of the contribution of the large hindgut,

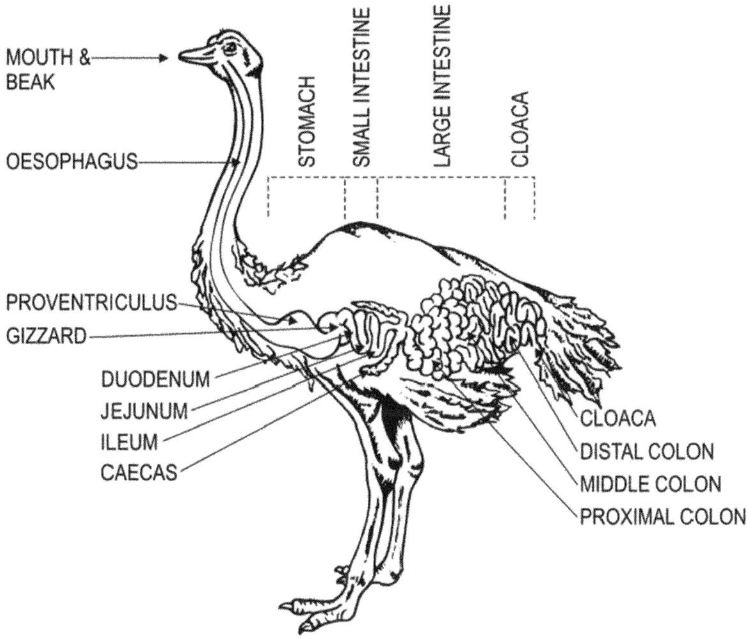

Fig. 5.1 The digestive system of the ostrich (Brand and Gous 2006a)

extract considerably more energy from the same feed than, e.g. pigs and poultry (Brand et al. 2000a, b). Concentrations of both amylase and trypsin in young chicks are low (Iji et al. 2003) and the application of exogenous digestive enzymes may therefore help to enhance digestion under intensive farming conditions.

5.3 Nutrient Requirements

The nutrient requirements of ostriches depend on the growth stage of the bird. Figure 5.2 illustrates the growth curve of South African Black ostriches under optimal feeding conditions.

The body composition of protein:fat ratio changes during the growing period and nutrient requirements will change accordingly. Dietary nutrient composition will also be dependent on feed intake, which in turn is dependent on the energy value of the feed that the birds consume. According to the values estimated from the Gompertz growth curve, the average daily gain of the birds initially increased and decreased again up to maturity (Kritzinger et al. 2009) (Table 5.1).

Under normal farming conditions, birds are fed different diets up to maturity. Table 5.2 illustrates the different commercial feeding periods for ostriches accompanied by the age of the bird and live mass, as well as the proposed energy value of the feed (Brand 2005, 2008).

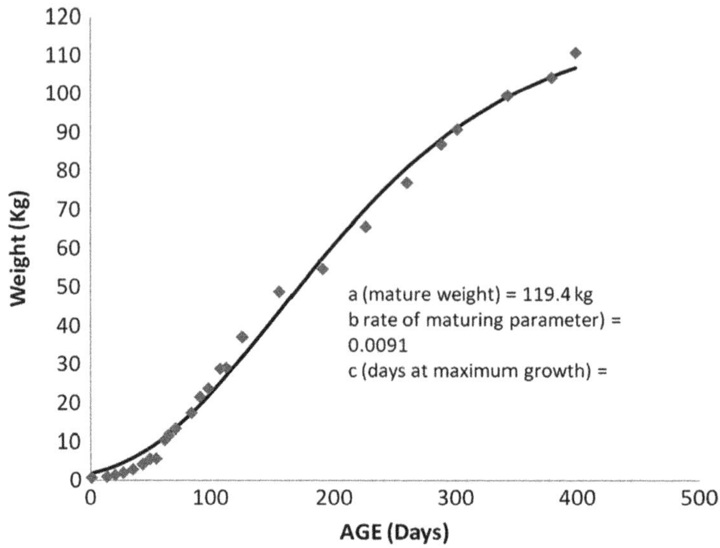

Fig. 5.2 Growth curve (calculated with the Gompertz equation) of South African Black ostriches under optimal feeding conditions (Kritzinger et al. 2009)

Table 5.1 Growth rate of ostriches on a free choice diet (Brand and Jordaan 2006)

Age (months)	Live mass (kg)	Growth rate (g/bird/day)
0–1	0.85–05.1	107
1–2	5.1–10.8	191
2–3	10.8–19.2	280
3–4	19.2–29.7	350
4–5	29.7–41.5	390
5–6	41.5–53.4	397
6–7	53.4–64.7	377
7–8	64.7–74.9	340
8–9	74.9–83.7	294
9–10	83.7–91.1	247
10–11	91.1–97.2	203
11–12	97.2–102.1	163
12–13	102.1–105.9	130
13–14	105.9–109.1	102

5.4 Feed Intake and Dietary Energy Requirements

5.4.1 Growing Birds

Table 5.3 illustrates the predicted feed intake of growing ostriches at three different dietary energy scenarios (i.e. 80%, 100% and 120% of proposed dietary energy values).

Table 5.2 Commercial feeding stages of growing ostriches (Brand 2005, 2008)

Feeding stages	Age (month)	Live mass (kg)	Growth rate (g/bird/day)	Proposed energy value of the feed (ME ostrich/kg feed)
Pre-starter	0–2	0.8–10	150	14.5
Starter	2–4.5	10–40	400	13.5
Grower	4.5–6.5	40–60	330	11.5
Finisher	6.5–10.5	60–90	250	9.5
Maintenance	10.5–12.0	90–100	200	8.5
Breeder	20+	110+	–	9.5

Table 5.3 Predicted feed intake (kg/bird/day) of growing ostriches at different ages up to maturity (Brand and Jordaan 2006)

Age (months)	80% of ME	Proposed ME	120% of ME
0–1	0.36	0.29	0.24
1–2	0.68	0.54	0.45
2–3	0.99	0.79	0.66
3–4	1.38	1.10	0.92
4–5	1.72	1.38	1.15
5–6	2.05	1.64	1.36
6–7	2.53	2.03	1.69
7–8	2.77	2.21	1.84
8–9	2.95	2.36	1.97
9–10	3.09	2.47	2.06
10–11	3.19	2.55	2.13
11–12	3.27	2.61	2.18
12–13	3.32	2.66	2.21
13–14	3.36	2.69	2.24
14–15	3.39	2.71	2.26
15–16	3.41	2.73	2.28
16–17	3.43	2.74	2.29

Studies by Brand et al. (2000a, b, 2005, 2006) revealed that the dietary feed intake of ostriches is dependent on the dietary energy value of the feed and feed intake decreases with higher dietary energy values.

5.4.2 *Breeding Birds*

In a study by Olivier et al. (2008, 2009) the ad libitum feed intake of mature breeding birds in clean breeding camps were found to be as high as 3.7 kg per bird per day (ME ostrich content from 8.0 up to 11.5 MJ/kg feed and a CP content of 12.8). Normally breeding birds in open veld camps are fed an amount of 2.5 kg per bird per day (ME ostrich of 9.5 MJ/kg feed and 12–14% CP). The energy level of the diet will determine the amount of feed needed to fulfil the requirements of the birds. With high density, feed daily intake may be lower and with low density, feed intake may be higher. Under grazing conditions and when breeder birds are kept in the South African veld they are normally fed 1.8–2.0 kg concentrated diet per bird per day. Studies by Brand et al. (2002, 2005) revealed that a dietary energy intake of less than 22 MJ ME ostriches bird per day will suppress egg production of female breeders and lead to a reduction in body mass.

5.5 Protein and Amino Acids Requirements

The protein and essential amino acid requirements of ostriches at different growth stages are presented in Table 5.4 (Du Preez 1991; Du Preez et al. 1992; Smith et al. 1995; Cilliers et al. 1998).

The exact dietary amino acid requirements of the bird will, however, be dependent on the feed intake of the bird as well as its growth rate. Commercial guidelines for ostrich feeds in South Africa are presented in Table 5.5 (Department of Agriculture (DOA) 2001). These guidelines are regulated by the Animal Feed Act of South Africa (Act 36 of 1947).

5.6 Mineral and Vitamin Requirements

The mineral and vitamin requirements of ostriches are calculated from values determined for other species. Commercial vitamin and mineral premixes are available for the different growth stages of ostriches. The minimum calcium (Ca)

Table 5.4 Predicted mean dry matter intake with accompanied protein and amino acid requirements for ostriches calculated from values published by du Preez (1991), du Preez et al. (1992), Smith et al. (1995) and Cilliers et al. (1998)

Parameter predicted	Production stage				
	Pre-starter	Starter	Grower	Finisher	Maintenance
Live mass (kg)	0.85–10.0	10–40	40–60	60–90	90–120
Age (months)	0–2	2–5	5–7	7–10	10–20
Feed intake (g/day)	275	875	1,603	1,915	2,440
Protein (g/100 g feed)	22.89	19.72	14.71	12.15	6.92
Lysine (g/100 g feed)	1.10	1.02	0.84	0.79	0.58
Methionine (g/100 g feed)	0.33	0.33	0.29	0.28	0.24
Cystine (g/100 g feed)	0.23	0.22	0.18	0.17	0.14
Total SAA (g/100 g feed)	0.56	0.55	0.47	0.45	0.38
Threonine (g/100 g feed)	0.63	0.59	0.49	0.47	0.36
Arginine (g/100 g feed)	0.97	0.93	0.80	0.78	0.63
Leucine (g/100 g feed)	1.38	1.24	0.99	0.88	0.59
Isoleucine (g/100 g feed)	0.70	0.65	0.54	0.51	0.38
Valine (g/100 g feed)	0.74	0.69	0.57	0.53	0.36
Histidine (g/100 g feed)	0.40	0.43	0.40	0.40	0.37
Phenylalanine (g/100 g feed)	0.85	0.79	0.65	0.61	0.45
Tyrosine (g/100 g feed)	0.45	0.44	0.38	0.38	0.31
Phenylalanine and tyrosine (g/100 g feed)	1.30	1.23	1.03	0.99	0.76

Table 5.5 Commercial guidelines (as fed) for the composition of ostrich feed (DOA 2001)

Feed type	Minimum crude protein (g/kg)	Minimum lysine (g/kg)	Maximum moisture (g/kg)	Minimum crude fat (g/kg)	Maximum crude fibre (g/kg)	Calcium		Minimum phosphorus (g/kg)
						Min (g/kg)	Max (g/kg)	
Pre-starter	190	10	120	25	100	12	15	6
Starter	170	9	120	25	135	12	15	6
Grower	150	7.5	120	25	175	10	16	5
Finisher	120	5.5	120	25	225	9	18	5
Slaughter	100	4	120	25	250	8	18	5
Maintenance	100	3	120	25	300	8	18	5
Breeder	120	5.8	120	25	240	20	30	5

requirements for growing birds vary between 0.8% and 1.2%, with maximum dietary levels of 1.5% and 1.8%, while phosphorus (P) requirements vary between 0.5 and 0.6%. Minimum/maximum Ca requirements of breeding birds varied between 2.0% and 3.0% (Department of Agriculture (DOA) 2001). Sometimes the practise to provide additional calcium in the form of oyster shell on an *ad libitum* basis to breeding females is followed.

Recent research (Kruger et al. 2008a) revealed a higher salt requirement for ostriches than that was previously recommended. This research indicated minimum requirements of 1% salt in diets of ostriches.

5.7 Water Requirements

In a study by Degen et al. (1991) it was found that the ratio of water consumption to dry matter intake (DMI) remained relatively constant, at approximately 2.3 l water/kg DMI. If ostriches are free-ranging, they consume juicy plants and therefore they seldom need to drink water (Berry and Louw 1982). Water intake will be affected by climate, pasture availability as well as feed intake and feed composition. Clean, cool drinking water should always be available.

5.8 Feed Processing

In an experiment by Kruger et al. (2008b) it was found that newly hatched ostrich chickens preferred a meal diet compared to a pelleted or extruded diet. In general, the following physical forms for ostrich diets are recommended (Brand 2005) (Table 5.6).

The roughage source (namely lucerne) should be milled through a 4.5-mm sieve size when used in pre-starter and starter diets. When other diets are provided as meal, it should be milled at least through a 6-mm sieve size for chickens and 12-mm sieve size for adult birds (Brand 2005). It has been found that pelleting improves the feed conversion ratio of birds by 10–15% (Brand et al. 2006). Impaction seems to be one of the main causes of mortality in ostrich chicks (Smith 1993) and therefore special attention should be given to feed processing.

Table 5.6 The recommended physical form of diets provided to birds at different production stages

Diet for production stage	Physical form of the diet
Pre-starter	Meal
Starter	Crumbs
Grower	Pellets (6 mm)
Finisher	Pellets (6 mm)
Maintenance	Pellets (6 mm)
Breeder	Pellets (6 mm)

5.9 Feeding Behaviour of Chickens

The mortality rate among ostrich chicks (refer Chaps. 4 and 9) is high (average mortality rate of 40%) (Kruger 2007). Apart from certain stress and health factors, mortalities may also be related to bad feeding behaviour because of the low ingestion of feed after the yolk sack has been absorbed. Several practical studies (Kruger 2007; Janse van Vuuren 2008) were therefore performed at the Kromme Rhee ostrich experimental farm near Stellenbosch in South Africa to determine the feeding behaviour of newly hatched ostrich chicks. Under intensive housing conditions, pen densities of more than 15 newly hatched birds per 4 m^2 suppressed feed intakes and performance of the chicks. Janse van Vuuren (2008) studied the effect of the presence of the breeding pair on the feeding behaviour of newly hatched chicks. The higher survival rate (82% versus 50%) of the chicks in the presence of their parents clearly illustrated the detrimental effect of intensive housing conditions on feeding behaviour. The provision of additional green material to chicks had a positive effect on feeding behaviour and activity and decreased the ingestion of soil. Feed intake and the growth rate of chicks fed on dry feed only were, however, higher than that of their counterparts receiving the additional green chopped lucerne (Janse van Vuuren 2008). In a further study by Janse van Vuuren et al. (2007a) ostrich chicks were exposed to a range of coloured flags while their pecking behaviour was observed. Although most of the pecks were at the green flags, successive studies with coloured feed do not reveal similar results. In other studies by both Kruger et al. (2008c) and Janse van Vuuren (2008, 2007a), no specific attraction of green coloured feed by newly hatched chickens were observed. It was concluded by Brand et al. (2008a, b) that the light intensity inside the chicken house may have affected the results obtained and that the results outside in bright sunshine may be different. In another study (Kruger et al. 2008d), ostrich chicks were fed diets with four different flavours namely sweet, salty, bitter, sour and a control diet in a free choice situation. The percentage intake from the artificially flavoured feeds were, respectively, 34.0% (salty), 17.9% (sweet), 17.1% (control), 15.7% (bitter) and 15.4% (sour). The study indicated that chicks with no previous exposure to feed have a preference for salty feed. Histological samples taken from the tongue, mouth cavity palate and upper part of the oesophagus of a 2-month-old chick, as well as an adult ostrich (12 months of age) showed that no conventional taste buds were present (Brand et al. 2008a, b). In two successive studies (Janse van Vuuren 2008), ostrich chicks were provided with diets with a meat, seafood, citrus, aniseed, lucerne, mint or control diet flavours in a free choice situation. Average intake of the birds was 18.6% (seafood flavour), 18.3% (control diet), 15.3% (meat flavour), 13.7% (aniseed flavour), 12.5% (citrus flavour), 12.0% (mint flavour) and 9.7% (lucerne flavour) in the first study. In the second study, average intake was 22.3% (seafood flavour), 19.8% (control diet), 12.4% (meat flavour), 12.6% (aniseed flavour), 6.6% (citrus flavour), 13.7% (mint flavour) and 12.7% (lucerne flavour).

Janse van Vuuren et al. (2007b) also investigated the provision of artificial high-intensity light on feed bowls on the feed intake of newly hatched ostrich chickens.

This increased the feed intake of the chickens by ca. 33%. The feed intake of birds from bowls equipped with green lights (48.2 g/day) were higher than that from bowls equipped with white lights (33.3 g/day) or a control (12.5 g/day) bowl with normal light.

A review by Glatz and Miao (2008) recommends a number of methods that can be used to stimulate chicks to commence eating; for example, (1) placing young chicks with older chicks, (2) foster parenting, (3) provision of boiled infertile eggs, (4) stirring feed by hand at least 8 times a day, (5) the use of attractive colours and (6) the provision of insects such as live crickets and mealworm. According to Glatz (2000a) feed intake of ostrich chicks can be increased by making the feed more attractive, improving the palatability and stimulating feeding by frequent stirring of the feed or frequent operation of automatic feeders.

5.10 Grazing Ostriches

Ostriches are selective grazers and graze by stripping shrubs or pastures from the leaves by running the branches through their beaks (Van Niekerk 1995). Birds normally feed on leaves and short grass blades and stems longer than 15 cm may result in compaction in the proventriculus, intestinal blockage and sudden death of the bird (Kok 1980). Ostriches are raised either extensively (birds are totally dependent on natural veld or cultivated pasture), semi-intensively (birds graze on veld or cultivated pasture and receive a feed concentrate as supplement) or intensively (a full balanced feed is provided) (Brand and Gous 2006a). Under natural grazing conditions in the veld, the average carrying capacity is about 10–12 ha per bird (Osterhoff 1979). The most common cultivated pasture used for grazing ostriches in South Africa is lucerne.

The normal stocking density of ostriches (refer Chap. 10) on lucerne is estimated to be approximately 6.5–12 slaughter birds per hectare (Nel 1993). Recent studies by Strydom et al. (2008) showed that ostriches may be kept at a stocking density of 15 birds per hectare on irrigated lucerne pasture provided that sufficient supplementary feed is provided to the birds. A previous study by Strydom et al. (2007) revealed that a concentrated supplement of 500 up to 1,000 g is necessary when raising ostriches on lucerne pasture for commercial purposes. Generally, considering the formulation of supplementary feeds for birds on lucerne pasture in South Africa, the assumed proportion of intake from pasture is 0% (pre-starter phase), 30% (starter phase), 50% (grower phase) and 70% (finisher phase). Farrell et al. (2000) found that supplementation levels below 70% of the total diet had a detrimental effect on growth. It is, however, true that both the quality and type of the pasture (which will determine to a great extent pasture and nutrient intake) and the quality or nutrient density of the supplement will determine the exact ratio of pasture to concentrate. It is important in feed formulation to make sure that the birds at the different growth stages will be able to consume enough nutrients from the two feeds sources to fulfil their nutrient requirements.

Overall, birds in captivity perform well on pastures dominated by any type of legume, for example, lucerne, clover, medics and seradella. Birds also perform well on broad-leaved plants such as canola or even salt bush (*Atriplex numularia*). Farrell et al. (2000) also concluded that pasture may be a viable economic alternative for ostrich producers compared to full feed on commercial diets. Pastures dominated by grass should also be avoided due to the possibility of impaction.

5.11 Nutrition and Welfare Implications

There is still a limited understanding of nutrition and nutritional physiology of ostriches, and its role in a production system. The partial understanding may result in an inability to provide the ostriches with their specific nutritional needs. When this happens, welfare consequences relating to imbalances, deficiencies and toxicosis may arise (Huchzermeyer 1998).

On the basis of the limited knowledge of nutrition and ostrich welfare, a number of focus areas are discussed as observed during extension to farmers experiencing nutrition-related production problems, which will have welfare implications. These are only observations, as research into welfare parameters has not been carried out to elucidate the potential effect on the ostrich. It must be further emphasised that these conditions are rare because of the impact poor nutrition may have on economical production parameters. The establishment of a stable sustainable industry over the years in South Africa has further reduced welfare problems through good stockmanship and the industry's self-controlled codes of practice. This may differ in the northern hemisphere production area or developing regions where climatic and geographic conditions and newcomer schemes may impact substantially on nutrition requirements or provision thereof (Deeming 1999). The basic principles, however, will remain the same.

The impact of diseases on the welfare of ostriches will be evaluated against the five freedoms that are considered to be the criteria for welfare determination. They are:

- Freedom from thirst, hunger and malnutrition (i.e. to sustain health and vitality)
- Freedom from discomfort, pain and injury
- Freedom from disease (i.e. also diseases induced or exacerbated by management)
- Freedom from fear and distress (i.e. protection from predators)
- Freedom to express most normal behaviour.

5.11.1 Feed and Water Management

Feed flow management is crucial to any livestock production system. Even when the system is extensive, with reliance on nature, all producers have to plan within the limitations of the future risks within the system (e.g. droughts). Their risk mitigation is focused on the carrying capacity of the planted or natural vegetation (crops) while assuring a build-up of additional grazing and improving the health and reproduction of the grazing vegetation as a resource. The economy of an intensive ostrich production system relies on the availability of feed, since feed is a high input cost (as high as 75% of total costs). Natural vegetation normally does not fulfil these requirements and most farmers in South Africa for example rely on irrigated or cultivated dry land pastures. Grazing of natural veld in the traditional ostrich production areas is also counter productive as overgrazing of natural vegetation rapidly occurs which leads to desertification and erosion. High reliance is therefore placed on fodder (alfalfa/lucerne or other legumes, or cultivated pastures) production for grazing and storage. As ostriches frequently receive total mixed rations they need to have access to clean, fresh potable water. Again, within an arid region and during high-temperature summer months, a high volume of cold water is required, which needs to be supplied on a constant basis. During the winter months, water troughs can be frozen and therefore management should address these issues to prevent birds going thirsty.

5.11.2 Stockmanship

Highly intensive ostrich systems usually have high bird numbers and extensive systems have low numbers of birds (refer Chaps. 6 and 9). Whatever the form of management or system, it is the responsibility of the stockman to ensure the bird's health and welfare. The stockman must consider the basic anatomical, physiological and behavioural needs irrespective of the system. Stockmen need to have a keen sense of observation to ensure rapid action to address any condition or situation which will affect the welfare of birds.

5.11.3 Feed and Water Intake

Numerous previous recommendations are to withhold feed and water from chicks to ensure adequate absorption of the yolk sac. This must be avoided, and chicks should have access to both feed and water from day 1 after hatching (Huchzermeyer 1998). Feed intake norms for starter, grower and finisher birds are well described by current research and producers should make sure that the feed availability are kept up to these norms. However, around 12 weeks of age the feed intake of ostrich chicks increases significantly and many chick growers do not take this into their feed flow calculations. The high cost of complete feed may be another limiting factor when a high-feed usage is seen in older birds. Ostriches must never be without feed as it will negatively affect production. As ostriches are extremely adaptive it would seem that they can cope with a lowered intake of feed, but this is a transgression of a key welfare aspect. Water access should be carefully managed.

There is no negative impact on production if water is withheld for a short period during the day. This may be needed during extremely high temperatures as some birds will over-imbibe water and keep other birds from gaining access to water. Access to cool water can then be given during the early mornings and late afternoons while restricting it during midday. As arid-adapted birds, this will not have a negative impact on health or welfare (Deeming 1999).

5.11.4 Feeding Space

Young birds have a high tolerance to each other at feed troughs and do not show aggressive behaviour towards their peers. They, however, do compete for feed and drinking space by pushing forward towards the feed or water trough. If insufficient space is available and feed is restricted in some manner, the weaker smaller birds do not have sufficient feed intake. Older grower ostriches (6 months) may have short skirmishes around feed troughs even with sufficient feeding space available.

5.11.5 Feeding Times and Conditions

Ostriches in the hot climate of South Africa have a high feed intake during the early mid-morning feeding time and again in the mid-afternoon. Birds need to have exposure to the optimum conditioned feed during these times. The feed should be fresh and provided at regular intervals throughout the day taking into consideration the behavioural feeding times, with less during midday in hot climates. Feeding during high day temperatures will also affect the composition of the feed (e.g. high temperatures will denature vitamins and lead to fat oxidation).

5.11.6 Feed and Water Quality

Feed and water quality are major contributors to intake of nutrients, assimilation for growth and physiological parameters. Anti-nutritional factors in feed will affect not only the physiological status (immunity suppression), but also production parameters. Feed microbiological status can also affect the health of birds as they are overly challenged by bacterial contaminants in the feed. Mouldy feed seems to have a deterring taste and birds will refuse feed in troughs with mouldy complete feeds. Some raw materials in the manufacturing of feeds such as molasses (strong smell) and canola oil cake (bitter taste) have an influence on acceptance of the feed if the birds are not used to it. The same holds true for feed with a high oil content, which can become rancid in high temperatures during summer. The rancidity will not only affect intake but may also result in gastric ulcers or deficiencies due to the high requirement of anti-oxidative vitamins and buffers.

5.11.7 Feed Supplementation

Many farmers supplement ostriches based on observed (perceived or true) deficiencies with a range of supplements. This is not needed if a balanced ration is given based on the physiological performance needs and management of the production system. Any wrongly diagnosed deficiency will have an impact on the welfare of the birds as an over-supplementation can result in a deficiency of another nutrient or a pathological condition. One such supplement is selenium, which, if incorrectly supplemented can result in acute toxicosis.

5.12 Nutritional-Related Diseases

5.12.1 Deficiencies and Imbalances

Deficiencies are seen when the physiological and production needs of the birds are not supplied or met due to a number of reasons (Deeming 1999; Huchzermeyer 1998). Stockmen may underestimate the need of the birds during certain age and production stages and will therefore provide feed that is not formulated correctly. An animal or bird factor which may play a role is the inability to assimilate or absorb the nutrients which are provided in sufficient quantities. Here stockmanship plays an important role to rapidly evaluate, identify and rectify the problem.

5.12.2 Calcium/Phosphorus

Dislocations of joints, spontaneous fractures of long bones, bending of long bones, thickening of the ends of long bones or twisting of long bones are frequent due to imbalances or deficiencies in macro- or micro-minerals. A common mistake by producers is to do on-farm mixing with rations formulated by a non-nutritionist. These are often unbalanced. Another cause for macro- and micro-mineral related problems is the omission of premixes or concentrates (deliberately or accidentally), with significant impact on the bird's welfare. Again the birds' ability to compensate and adapt is remarkable and they will continue to thrive with leg deformities. It is only when there is complete joint luxation or fracture that they will be unable to feed. These deformities may lead to ulcers and sores on joints or padding as birds continue to struggle. Birds with thickening of the ends of the long bone often show pain and stress responses to these conditions. They tremble and are reluctant or refuse to stand and walk. They show signs of stress by stretching their long neck flat on the ground while flapping their wings and opening and closing their beaks. Birds with leg deformities seldom recover and should be euthanised as soon as possible.

5.12.3 *Pica*

Another behavioural abnormality of chicks suffering from micro- and/or macro mineral deficiencies is pica. This results in the ingestion of foreign material such as sand or sticks, which damages the gizzard wall because of penetration injuries or has a corrosive action on the mucosa of the intestines. Another factor contributing to pica is the quality of fibre, as older chicks seem to prefer effective fibre of 8–15 mm in length.

5.12.4 *Vitamin E/Selenium*

A condition which is frequently seen in ostrich production units is a Vitamin E and/or selenium deficiency. This is observed in birds with a high energy and protein ration showing exceptional growth. It is also seen in birds with a high energy and protein ration given additional maize in an attempt to further improve their growth rate. Birds are reluctant to stand and cannot display normal movement or feeding behaviour. Birds suffering from lowered intake have a resulting lowered vitamin intake and develop deficiency signs of scaling and crusting of the skin. The crusts are infected with secondary bacteria which worsens the conditions. The chicks not only suffer from disease, but also tend to lose some of their eyesight and cannot ingest feed.

5.12.5 *Clostridial Overgrowth*

When birds are fed for production, various changes take place in the digestive tract to accommodate the highly nutritious rations. This is one aspect that has been paid little research attention. The diet change and resulting intestinal changes (physiological and intestinal flora) predispose the chicks to disease such as clostridial overgrowth. This results in damage to the intestinal wall and often ends in acute death. These conditions can be prevented by gradual feed changes and the elimination of stress factors which seem to trigger the overgrowth of flora in the intestinal track.

5.12.6 *Mycotoxins*

Mycotoxins are products of secondary metabolism of a number of species of fungi. They are toxic in small quantities. Fungi occur naturally in feeds or feed raw material and can significantly affect the health status and economic production parameters of an enterprise (Pasteiner 1998). The fungal metabolites affect the metabolic systems (lipid and carbohydrate metabolism, vitamin assimilation, etc.), endocrine system and the skeletal system. Symptoms and signs can vary due to amount, time and type of metabolite ingested. Low levels of mycotoxins will predispose the birds to infections through the impairment of the humoral and cellular immune response. It will also impair the ability to develop resistance through antibody production.

Apart from the metabolic effect, it is observed that the fungal metabolites have an additional effect on the palatability of the feed and therefore feed intake. Overall, mycotoxins have a negative impact on the immune system, metabolism and intake which increases the individuals risk to disease.

5.13 Concluding Remarks

The ostrich industry is still young compared to other livestock industries and a lot of research is needed to assure optimum feeding, health and welfare of the birds. Mortality amongst young chicks in intensive production systems is very high (30–40%) and although the primary causes are still unknown, special attention to the feeding and feeding behaviour of young chicks, especially under intensive raising conditions, is recommended. An overview is done in this chapter of the recent research on this important welfare aspect of ostrich chicks kept in an intensive production system. The chapter also gave a clear indication of the major nutritional needs of ostriches during the different stages of production. These requirements and other current research data are currently built into a mathematical optimisation model for ostriches, which will help to put the nutritional needs of ostriches on a more scientific base and prevent the over or under consumption of certain nutrients, especially with the variation in feed intake by the bird due to several dietary, genetic and environmental factors (Gous and Brand 2008). Practical guidelines on several nutrition and welfare related factors given in the chapter are based on practical observations in ostrich production systems in South Africa and should be kept in mind when farming with ostriches. The most observed nutrition-related diseases as observed during post mortem of birds at the specialist veterinary clinic in Oudtshoorn, South Africa, and during on farm consultation is also discussed and should be noticed to prevent repetition in an attempt to enhance the welfare of ostrich chicks.

References

Angel CR (1996) A review of ratite nutrition. Anim Feed Sci Technol 60:241–246

Berry HH, Louw GN (1982) Nutritional balance between grassland productivity and large herbivore demands in Etosha National park. Modoqua 13:141–150

Brand TS (2005) Ostrich nutrition technical brochure. Elsenburg Animal Production Institute, South Africa, pp 12 (In Afrikaans)

Brand TS (2008) Ostrich nutrition: a scientific approach. Sun Print, University of Stellenbosch, Stellenbosch, South Africa, pp 48 (In Afrikaans)
Brand TS, Cloete SWP (2007) Achievements of research in the field of ostrich production. Animal production and animal science worldwide. World Association for Animal Production, Waginingen Academic, Italy, pp 18–194
Brand TS, Gous RM (2006a) Feeding ostriches. In: Bels V (ed) Feeding domestic vertebrates. CAB International, UK, pp 136–155
Brand TS, Gous RM (2006b) Ostrich nutrition using simulation models to optimize ostrich feeding. Feed Technol 7:12–14
Brand TS, Jordaan JW (2006) Computer Program: optimal slaughter age for ostriches. Elsenburg Animal Production Institute, Western Cape Department of Agriculture, Private Bag X1, Elsenburg 7607, South Africa
Brand TS, Brabander D, Van Schalkwyk SJ, Pfister B, Hayes JP (2000a) The true metabolisable energy content of canola oilcake meal and full-fat canola seed for ostriches (Struthio camelus). Br Poult Sci 41:201–203
Brand TS, Salih M, Brand Z (2000b) Comparison of estimates of feed energy obtained from ostriches with estimates obtained from pig, poultry and ruminants. S Afr J Anim Sci 30:13–14
Brand Z, Brand TS, Brown CR (2002) The effect of dietary energy and protein levels on the production of breeding female ostriches. Br Poult Sci 44:598–606
Brand Z, Brand TS, Brown CR (2005) The effect of dietary energy and protein levels on body condition and production of breeding male ostriches. S Afr J Anim Sci 32:231–239
Brand TS, Aucamp BB, Kruger ACM (2006) The effect of pelleting on the diet utilization by ostriches. In: Proceedings of the 41st Congress of the South African Society Animal Science, Bloemfontein, South Africa, 3–6 Apr 2006, p 164
Brand TS, Janse van Vuuren M, Aucamp BB (2008a) Studies on the feeding behavior of ostrich chickens. In: Proceedings of the Ostrich Information day, Oudtshoorn, South Africa, June 2008
Brand TS, Kruger ACM, Aucamp BB (2008b) The effect of feeding management practices on the feed intake and production performance of newly hatched ostrich chickens. In: Proceedings of the Ostrich Information day, Oudtshoorn, South Africa, June 2008
Cilliers SC, Hayes JP, Chwalibog A, Sales J, Du Preez JJ (1998) Determination of energy, protein and amino acid requirements for maintenance and growth in ostriches. Anim Sci Technol 72: 283–293
Deeming DC (1999) The ostrich: biology, production and health. CABI, London
Degen AA, Kam M, Rosenstrauch A, Plovnic P (1991) Growth rate, total body volume, dry-matter intake and water consumption of domestic ostriches (Struthio camelus). Anim Prod 52:225–232
Department of Agriculture (DOA) (2001) Guidelines for the composition of animal feeds. Department of Agriculture, Pretoria, South Africa
Du Preez JJ (1991) Ostrich nutrition and management. In: Farrell DJ (ed) Recent advances in animal nutrition in Australia. University of New England, Armidale, pp 278–291
Du Preez JJ, Jarvis MJF, Capatos D, De Kock JA (1992) A note on growth curves for the ostrich (Struthio camelus). Anim Prod 54:150–152
EFG Software (2008) EFG broiler model. http://www.efgsoftware.net
Farrell DJ, Kent PB, Schermer M (2000) Ostriches, their nutritional needs under farming conditions. A report for the Rural Industries Research and Development Corporation, Australia
Glatz P (2000a) A benchmark study of husbandry, transport, lairage and slaughter methods to improve skin quality of ratites, vol 1. Ostriches. A report for the Rural Industries Research and Development Corporation, Australia
Glatz P, Miao Z (2008) Reducing mortality rates in ostrich chicks. Rural Industries Research and Development Corporation, Australia, pp 1–27
Gous RM, Brand TS (2008) Simulation models used for determining food intake and growth of ostriches: an overview. Aust J Exp Agric 48:1266–1269
Huchzermeyer FW (1998) Diseases of ostriches and other ratites. Agricultural Research Council, South Africa

Iji PA, Van der Walt JG, Brand TS, Boomker EA, Booyse D (2003) Development of the digestive function in the ostrich. Arch Anim Nutr 57:217–228
Janse van Vuuren M (2008) Factors that will effect the survival of ostrich (Struthio camelus) chicks. Nelson Mandela Metropolitan University, Saasveld, George, South Africa (In Afrikaans)
Janse van Vuuren M, Brand TS, Aucamp BB (2007a) Colour preferences in ostrich (Struthio camelus) chickens. In: Proceedings South African Society Animal Science. Warmbaths, South Africa
Janse van Vuuren M, Brand TS, Aucamp BB (2007b). The effect of direct artificial light of high intensity on feed bowls on feed intake, growth and behavior of newly hatched ostrich (Struthio camelus) chickens. In: Proceedings South African Society Animal Science. Warmbaths, South Africa
Jordaan JW, Brand TS, Bhiya C, Aucamp BB (2008) An evaluation of slaughter age on the profitability of intensive slaughter ostrich production. Aust J Exp Agric 48:916–920
Kok OB (1980) Feed intake of ostriches in the Namib-Naukluft park, Suidwes-Afrika. Modoqua 12:155–161 (In Afrikaans)
Kritzinger W, Brand TS, Hoffman LC, Mellet FD, Aucamp BB (2009) Growth and body composition of ostriches under optimal feeding conditions. In: Proceedings SASAS Congress, Bergville, South Africa, 28–30 July 2009
Kruger ACM (2007) The effect of different management practices on the feed intake and growth rate of ostrich chicks. M Tech-thesis, Nelson Mandela Metropolitan University, Saasveld, George, South Africa
Kruger ACM, Brand TS, Aucamp BB (2008a) The effect of different dietary salt levels on the feed intake and growth of ostrich chicks. In: Proceedings of Annual Conference of the South African Society for Agricultural Technology, West Coast, South Africa, 16–19 Sept 2008
Kruger ACM, Brand TS, Aucamp BB (2008b) The effect of feed processing and restriction of water availability on the intake and growth of ostrich chicks. In: Proceedings of 26th Annual Congress of the South African Society for Agricultural Technology, West Coast, South Africa, 16–19 Sept 2008
Kruger ACM, Brand TS, Aucamp BB (2008c) The effect of artificially coloured mash on the feed intake of ostrich chicks. In: Proceedings of 26th Annual Congress of the South African Society for Agricultural Technology, West Coast, South Africa, 16–19 Sept 2008
Kruger ACM, Brand TS, Aucamp BB (2008d) The effect of artificially flavoured mash on the feed intake of ostrich chicks. In: Proceedings of 26th Annual Congress of the South African Society for Agricultural Technology, West Coast, South Africa, 16–19 Sept 2008
Nel JC (1993) Pasture utilization by ostriches. Report on the ostrich industry in South Africa. Oudtshoorn Development Centre, Oudtshoorn, South Africa
Nizza A, Di Meo C (2000) Determination of the apparent digestibility coefficients in 6-, 12- and 18-week-old ostriches. Br Poult Sci 41:518–520
Olivier TR, Brand TS, Brand Z (2008) Influence of dietary energy levels on production of breeding ostriches. In: Proceedings of the World Congress on Animal Production, Cape Town, 26 Nov 2008
Olivier TR, Brand TS, Brand Z (2009) Production and the effect of dietary energy level on the feed intake of breeding ostriches. In: Proceedings South African Society of Animal Science Congress, Bergville, South Africa
Osterhoff DR (1979) Ostrich farming in South Africa. World Rev Anim Prod 15:19–30
Pasteiner S (1998) Mycotoxins and Animal Husbandry. Biomin, Industriestrasse, Herzogenburg, Austria
Salih ME, Brand TS, Van Schalkwyk SJ, Blood JR, Pfister B, Brand Z, Akbay R (1998) The effect of dietary fibre level on the production of growing ostriches. In: Proceedings of the Second International Ratite Congress, Oudtshoorn, South Africa, pp 31–37
Smith CA (1993) Research roundup: ostrich chick survival presents challenge. J Am Vet Med Assoc 203:637–641
Smith WA, Cilliers SC, Mellet FD, Van Schalkwyk SC (1995) Nutrient requirements and feedstuff values in ostrich production. Feed Comp(Sept):2–29

South African Ostrich Business Chamber (SAOBC), Co-operation Street, Oudtshoorn 6620, South Africa
Strydom M, Brand TS, Aucamp BB, Van Heerden JCM (2007) The effect of supplementary feeding on the production of grazing ostriches (Struthio camelus). In: Proceedings of the South African Society of Animal Science, Warmbaths, South Africa
Strydom M, Brand TS, Aucamp BB, Van Heerden JCM (2008) Effect of two different levels of supplementary feed and two different stocking rates on the production of ostriches grazing irrigated lucerne pasture. In: Proceedings of the World Congress on Animal Production, Cape Town, 26 Nov 2008
Swart D (2008) Studies on the hatching, growth and energy metabolism in the ostrich chick (Struthio camelus). PhD thesis, University of Stellenbosch, Stellenbosch
Van der Walt JG, Iji PA, Brand TS, Boomker EA, Booyse D (2008) Early development of the digestive function in the ostrich. In: Proceedings International Herbivore Symposium, Mexico
Van Niekerk BDH (1995) The science and practice of ostrich nutrition. In: Proceedings of the AFMA Conference, Pretoria, June, pp 1–8

Chapter 6
Welfare Issues Associated with Ratite Husbandry Practices

P.C. Glatz

Abstract This chapter examines some of the methods being used to assess welfare in ratites, identifies the welfare issues associated with declawing and highlights concerns with stockperson skills and brooder house management.

Partial amputation of the toes of ratites has welfare implications. It causes ratites acute pain; they become flatfooted and change their gait and they can slip and fall over in wet paddocks and handling areas. However, the blunting of the claws reduces the bird's ability to deliver kicks and injuries to other birds during aggressive encounters and improves skin quality in the flock and reduces the potential for handlers suffering injuries.

The role of handlers in ratite farming is examined. While there has been little research undertaken in the ratite industry on skills of stockpersons, work with other livestock shows that it is essential that animal handlers have the skills to look after their stock and develop a good affinity with their birds.

Livestock which are housed in facilities with good air quality grow faster and consume more feed than animals exposed to poor air quality. Improving air quality in ratite brooding facilities could improve production and provide a better working environment for farm employees.

It is clear that declawing of ratites, stockperson skills and housing have a major impact on welfare. The accreditation of persons practising declawing, the training of stockpersons in ratite handling and improvement in air quality in brooding and rearing facilities would result in a significant improvement in ratite welfare.

6.1 Introduction

There has been a concern expressed for ratites that they are confined under intensive housing conditions (Uhart et al. 2006). Consequently Welfare Codes of Practice for farmed ratites have been developed in some countries to define the minimum

P.C. Glatz
SARDI, Roseworthy Campus, University of Adelaide, Roseworthy, SA 5371, Australia
e-mail: phil.glatz@sa.gov.au

P.C. Glatz et al. (eds.), *The Welfare of Farmed Ratites*, Animal Welfare 11,
DOI 10.1007/978-3-642-19297-5_6, © Springer-Verlag Berlin Heidelberg 2011

standards required to maintain bird welfare (Standing Committee on Agriculture and Resource Management 2003; Animal Welfare Advisory Committee 1998). In addition, laws exist in some countries to prevent people from being cruel to birds. For example, in Australia, the Office of the Queensland Parliamentary Counsel (2009) has a law which penalises people who are cruel to animals. The Food and Agriculture Organisation of the United Nations places importance on practices that lead to benefits for animals and supports implementation of practices that improve welfare (Food and Agriculture Organization of the United Nations 2009).

In the European Union (EU) there was a concern that ostriches, which thrive in hot dry climates, would find it difficult to live in a cold winter climate. The EU Council Directive 98/58 "Concerning the protection of animals kept for farming purposes" prohibits the breeding of animals if it causes detrimental effects on their health or welfare (European Communities 1998). The EU also issued a recommendation that ostrich farms should only be located in areas where environmental conditions allow birds to be kept outside most of the day in any season, to satisfy their need for exercise and grazing.

In many countries, various livestock industries have established quality assurance programs and animal welfare is included in audits of farms (Barnett et al. 2001). The audits have occurred in response to consumer demand to ensure animal products attain certain safety, environmental and welfare standards.

In other intensive livestock industries (pigs and poultry), the major welfare issues are housing, amputation procedures, stockmanship, transport, processing and euthanasia (Barnett and Newman 1997; Barnett et al. 2001). These issues apply equally to ratites. This chapter examines some of the major contentious husbandry issues in the ratite industries with reference to methodology being used to assess welfare.

6.2 Definition of Welfare

There has been considerable debate about how animal welfare should be assessed (Fraser 2003; Sandøe et al. 2004) with many definitions provided for animal welfare (refer Chaps. 1 and 2). Here we adopt the following definition: the provision of good welfare for livestock means meeting high standards of husbandry, which includes care of animals, good housing, protection from the environment, maintaining good health, preventing disease, recognising and treating disease, providing good nutrition and good stockpersonship.

The variation in both the definition and methods of assessing welfare has resulted in considerable debate and disagreement on how welfare should be assessed and interpreted (Hemsworth and Coleman 1998).

A few of the many welfare definitions that have been proposed are as follows:

The Oxford English Dictionary associates welfare with "well-being; happiness; and thriving or successful progress in life".

Brambell et al. (1965) stated that welfare is a wide term that embraces both the physical and mental well-being of the animal. Attempts to evaluate welfare, therefore, must take into account the scientific evidence available concerning the feelings of animals that can be derived from their structure and function and also from their behaviour.

In the Saunders Comprehensive Veterinary Dictionary, animal welfare is defined as "the avoidance of abuse and exploitation of animals by humans by maintaining appropriate standards of accommodation, feeding and general care, the prevention and treatment of disease and the assurance of freedom from harassment, and unnecessary discomfort and pain".

Broom (1986) defined the welfare of an animal as "its state as regards its attempts to cope with its environment. This state includes how much it has to do to cope, the extent to which it is succeeding in or failing to cope, and its associated feelings. Welfare will vary over a continuum from very good to very poor and studies of welfare will be most effective if a wide range of measures is used".

There is a requirement for providing animals the five freedoms; freedom from hunger and thirst; freedom from discomfort; freedom from pain, injury and disease; freedom to express most normal behaviour and freedom from fear and distress (Anon 1992). It is also important that animals should be able to express most normal behaviours when kept in an intensive farming environment (Weeks and Nicol 2006) and that they should not suffer from pain, fear and distress.

In this chapter, emphasis is placed on assessing welfare of ratites based on their behaviour, particularly for declawed birds, their interaction with stockpersons and the ability of ratites to cope in a farming environment.

6.3 Assessment of Ratite Welfare

Brambell et al. (1965) recommended that all animals are entitled to good welfare and defined the basic freedoms referred to above. Progress is being made towards meeting the five freedoms in intensive ratite production (Glatz and Miao 2008).

In Europe, a project (LayWel 2006) developed a series of welfare assessment protocols for poultry. However, no such protocols have been specifically developed for assessing ratite welfare. The poultry welfare assessment system has an emphasis on scoring animals according to their health status, feather cover, injuries and behaviour. The welfare scores reflect how the bird is interacting with its environment. Current methods of assessing welfare in the ratite industry have concentrated on assessing the impacts of housing and husbandry on production and behaviour (Glatz and Miao 2008). However, a ratite skin quality audit (Glatz 2001a) used a scoring system to assess risk factors that lead to skin injuries. There is a significant welfare problem associated with skin injuries in farmed emus and ostriches. Injuries result in the loss of production, pain and skin infections in birds (Meyer et al. 2003; Glatz 2001a). The factors that lead to skin damage and poor welfare include aggressive interactions between birds from kicking, feather pecking and trampling. In addition panic in the flock often results in the fright response and birds attempt to escape from confined areas and become trapped in fences or injure themselves when they run into fences. The ratite skin audit (Glatz 2001a) also identified risk factors that lead to injuries during yarding, loading/unloading of birds onto trucks, transport and negative interactions between birds in the lairage. These issues are discussed in Chap. 10 in this volume.

To assess bird welfare it is important that observations and measurements are made on individual animals. For example, the poultry welfare quality project (LayWel 2006) used the welfare freedoms as the basis for assessing bird animal welfare and focused on four welfare categories. These included (1) injury, disease and pain; (2) hunger, thirst and productivity; (3) behaviour and (4) fear, stress and discomfort. Scores were given for a range of welfare risks in the various poultry production systems. The findings suggest that birds housed in more intensive systems are at a greater welfare risk and this may also apply to ratites that are confined to small paddocks.

Declawing using a heated blade is the most controversial husbandry practice in ratites and is used here as an example of the different approaches that have been used to appraise ratite welfare. One approach assessed welfare of declawed birds on the basis of their performance and skin quality (O'Malley and Snowden 1999; Glatz 2001b). The results from these studies showed that the husbandry practice of declawing had a positive impact on bird welfare by reducing injuries to other birds, but the initial impact from the surgery was to cause a reduction in growth.

Another approach used to determine the welfare impact of declawing was to assess the potential chronic pain in the toe stump by making an anatomical assessment of the incidence of neuromas (Lunam and Glatz 2000). The presence of neuromas is an indicator that the bird may be feeling chronic pain in the toe stump and is a negative emotion that birds may experience as discussed by Duncan and Fraser (1997).

The welfare state of birds has also been assessed using preference and behavioural demand tests (Dawkins 1980). For example, preferences of ostriches were evaluated by providing them a choice of an enriched or a barren environment and determining the time they spent in the chosen area (Christensen and Nielsen 2004). These tests determined if preferences are influenced by the animal's feelings and place a value on the bird's choice (Dawkins 1983) particularly when an evaluation is made of how hard the animal works to obtain their preference. In the case of ratites, preference testing of declawed birds versus control birds has not been evaluated in terms of the resources they will select and if there is a decline in the strength of their demand to access a facility or resource because they may be feeling pain from declawing.

A further approach that has been used to assess welfare of declawed ratites has been to determine if the operation caused a change in the locomotor behaviour (see Chap. 8) and other behaviour of birds (Glatz 2001b, 2010). However, Dawkins (2003) indicated that it is difficult to attribute actual pain in birds if certain behaviours are absent. More recently there has been a greater emphasis on behavioural indicators of poor coping such as fearfulness, aggression and stereotypies. For acute pain (such as in declawing) there is a need to assess performance of animals (Mellor et al. 2000), which was done in declawed ratites showing an initial decline in performance (O'Malley and Snowden 1999; Glatz 2001a). However, for assessment of chronic pain or similar long-term effects on the birds, it is clear that a range of welfare parameters should be assessed which may lead to varying opinions on bird welfare.

6.4 Key Welfare Issues in the Ratite Industry

The welfare issues in the ratite industries were previously reviewed (Glatz and Miao 2008) and have been addressed in greater depth in other chapters in this volume (see Chap. 2 on breeders, Chap. 4 on incubation and hatching, Chap. 10 on brooding, rearing systems, housing, transport and slaughter). In commercial poultry and pigs, the welfare issues, which have raised the most concern have included housing, stocking density and group size, leg weakness, restricted feeding and amputations (Barnett and Newman 1997; Barnett et al. 2001). Most of these issues are also a concern in the ratite industries although only limited information has been given to them in this volume. Four issues are discussed in greater depth in this chapter namely (1) declawing, (2) feather pecking, (3) stockmanship and (4) air quality in houses.

6.4.1 Declawing in Farmed Ratites

Approximately one-third of ostrich skins and one-half of emu skins are downgraded because of the presence of scars on the hide. Some of these scars are caused by claw abrasions (Glatz 2008). The hide damage in ratites occurs throughout the growing cycle. Birds are naturally aggressive towards each other and use their feet to kick at each other. This behaviour can become a problem in commercial situations. Farm managers have their flocks' declawed to blunt the claws enough so that kicking does not result in damage to the skin (Glatz 2008). Declawing also improves safety of farm workers by reducing the potential for injuries to workers during handling of birds. Anecdotal reports from the ratite industries indicate that adult birds tend to be more relaxed and less aggressive when they are declawed. Handlers feel more confident with the birds because they are not concerned with receiving claw damage from the bird and a better human bird relationship develops. Some ostriches hatch with a curled toe problem. Declawing of these birds prevents the need to cull them and allows the bird to walk and forage normally. In Australia declawing is practised on most emu farms and a few ostrich farms (Glatz 2008).

Declawing is carried out with day-old birds using a heated blade that cuts and cauterises the terminal segment of each toe (see Chap. 4). The operator holds the bird securely with its claw resting on the cutting plate. A foot pedal is then operated to bring the heated blade down onto the claw. The blade cuts through the claw in one motion. The heat of the blade seals off the cut thus preventing bleeding and

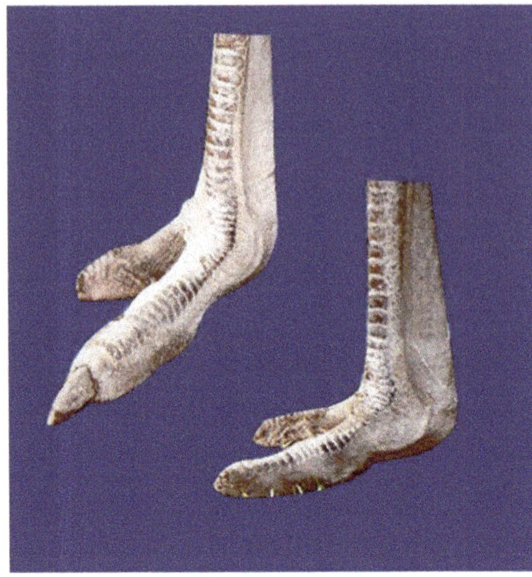

Fig. 6.1 Declawed and intact claw of ostrich

infection (Glatz 2008). The claw can regrow after it is cut if the toes are trimmed incorrectly. Declawing techniques, which retain the emus toe pad under the claw, will maximise the development of a pad of tissue at the tip of the declawed toe and provide a cushion at the end of the toe (Fig. 6.1). The development of this pad would appear to protect the tip of the toe against injury from rocks and hard surfaces (Glatz 2008). The best results can be achieved using a hot blade machine to remove the distal phalangeal joint, fitted with a convexed bottom or guide bar and operated to maximise the retention of pad tissue under the claw (O'Malley and Snowden 1999).

In the USA, Nova-Tech Engineering, Inc. (J. Sieben, pers. comm.) has developed an alternative method of claw removal for poultry, using microwaves to treat the claws. When the process is complete, the toes and claws remain intact; there is no blood loss or an open wound. Domestic poultry remain mobile and active, and they can consume feed and water sooner. The microwave claw treatment process disrupts the underlying germinal cell tissue, which the outer, harder, claw (keratin) is generated from. As a result, the claws stop growing and the original claws fall off in the first few weeks of the bird's life. However, there is no data available on the short and long-term welfare consequences of this procedure.

6.4.2 Declawing of Ostriches and Emus

The short and long-term effects of day-old declawing on the behaviour of ostriches in the first 8 h after the operation, at 1–2 weeks-of-age and at 12 months-of-age were examined in an on-farm study (Glatz 2010). The welfare issues associated with any amputation (e.g. declawing) can be divided into three phases. There is acute pain during and immediately after the operation (Grigor et al. 1995) until several days later (Lee and Craig 1990), sensory deprivation during the animal's life (Hughes and Michie 1982), and chronic pain from neuromas (Gentle 1986; Breward and Gentle 1985).

6.4.2.1 Behaviour of Day-Old Declawed Ostriches

In an on-farm study (Glatz 2010) declawed chicks engaged in fewer bouts of standing (Table 6.1) compared to control birds (not declawed) after the operation indicating the declawed birds may have been suffering acute pain and discomfort from the operation. In addition, declawed chicks had more bouts of sitting, fewer ($P = 0.11$) bouts of stepping on and being stepped on ($P = 0.08$) by other declawed birds and a reduction in other activities such as pecking at the environment, kantling and giving threats (Table 6.2). Declawing day-old chicks appears to change the behaviour of ostrich chicks and while it may reduce the potential for causing skin damage to other birds it is likely that the declawed birds were in acute pain.

Table 6.1 Effect of declawing on day-old ostrich chick behaviour (bouts) averaged over 1h intervals from 2000h–0800h on day of operation

Treatment	Eat	Drink	Preen self	Preen others	Pecked by other bird	Peck another bird
Control	25.1	5.2	37.4	1.8	5.3	9.1
Declaw	23.9	3.8	34.5	1.9	8.8	11.5
P	0.617	0.085	0.302	0.982	0.055	0.711
LSD ($P = 0.05$)	NS	NS	NS	NS	NS	NS

NS not significant in analysis of variance, *P* is probability value from analysis of variance

Table 6.2 Effect of declawing on ostrich chick behaviour (bouts) averaged over 1h intervals from 2000h–0800h on the day of the operation

Treatment	Sit (bouts)	Stand (bouts)	Step on	Stepped on	Peck environment	kantling	Threats
Control	68.7a	69.1a	6.6a	6.4a	23.6a	7.3a	15.2a
Declaw	57.5b	57.8b	4.7a	4.0a	13.7b	2.8b	3.5b
P	0.017	0.017	0.112	0.085	0.005	0.001	0.001
LSD ($P = 0.05$)	8.9	9.1	2.4 ($P = 0.11$)	2.7 ($P = 0.08$)	6.6	2.5	5.8

Means within columns followed by the same letter are not significantly at $P = 0.05$; *LSD* least significant difference, *P* probability value from analysis of variance

6.4.2.2 Behaviour of Week-Old Declawed Ostrich Chicks

In week-old ostriches comparison of inactive, ingestive, grooming, aggressive and locomotor behaviours has been made in declawed birds and control birds in a brooder area and in an outdoor run (Glatz 2010). In the brooder area there were no differences in behaviour of birds (Table 6.3). However, in the outdoor run (where birds had greater ability to express their locomotor behaviours) declawed birds engaged in more bouts of sitting and standing. This suggests that birds were still feeling discomfort in the toe stumps as pain from amputations can last for some time after the operation in other bird species (Grigor et al. 1995; Lee and Craig 1990). It is likely that neuromas are forming in the first week after the operation (Gentle 1986; Breward and Gentle 1985; Lunam and Glatz 2000), which is implicated as the cause of chronic pain in birds (Breward and Gentle 1985; Gentle 1986).

Biomechanical studies on the emu's walking behaviour indicate that the gait of birds changes after declawing and they became flat footed (Lunam and Glatz 2000). Field reports from the ostrich industry (Glatz 2001a) also indicate that declawed ostriches tend to lose their balance and slip. It was also observed in the on-farm behaviour study (Glatz 2010) that declawed ostriches fall over ($P = 0.08$) and

Table 6.3 Effect of declawing on bouts that 10 ostrich chicks (1–2 weeks-of-age) engaged in inactive, ingestive, grooming, aggressive and locomotor behaviours averaged over 1h intervals for 8h (0800–0400h) in the indoor brooder area and outside run

Variable	Brooder				Outside run			
Inactive	Declaw	Control	*P*	LSD	Declaw	Control	*P*	LSD
Sit	9.8	8.5	0.36	NS	3.3a	0.7b	0.003	1.6
Stand	9.4	7.9	0.36	NS	3.2a	0.8b	0.0004	1.6
Ingestive								
Eat	10.9	12.7	0.62	NS	0	0	–	–
Drink	4.8	5.4	0.66	NS	0	0	–	–
Grooming and other behaviours								
Posture change	1.8	1.4	0.28	NS	0.04	0	0.32	NS
Preen self	6.0	4.9	0.39	NS	0.28	0.2	0.58	NS
Head scratch	0.4	0.4	0.67	NS	0	0.04	0.32	NS
Stretch	0.76	0.72	0.89	NS	0.6	0.24	0.15	NS
Peck environment	45.0	55.3	0.19	NS	30.5	20.3	0.16	NS
kantling	0	0	–	–	0	0.04	0.89	NS
Aggressive								
Runaway	4.6	4.6	0.96	NS	5.2	3.9	0.41	NS
Peck	1.7	3.3	0.76	NS	0.24	0.10	0.27	NS
Pecked	1.9	2.3	0.23	NS	0.32	0.12	0.22	NS
Step on	0.24	0.12	0.33	NS	0	0	–	–
Stepped on	0.36	0.16	0.29	NS	0	0	–	–
Locomotor								
Walk	61.1	60.4	0.91	NS	32.9	18.9	0.05	NS
Pace	1.6	0.6	0.15	NS	0.5	0	0.05	NS
Bump	0.2	0.1	0.16	NS	0.5	0.2	0.19	NS
Fall	0.3	0.04	0.12	NS	0.32	0.08	0.08	NS

Means within rows followed by the same letter are not significantly at $P = 0.05$; *LSD* least significant difference; *P* probability value from analysis of variance bump into each other ($P = 0.05$) more than control birds (Table 6.3). Declawed ostriches at 1 week of age had greater difficulty in their locomotor behaviour probably due to a change in their gait and balance when walking, pacing and running and perhaps due to continuing pain as a result of the operation.

6.4.2.3 Behaviour of 1-Year-Old Declawed Ostriches

There were no differences reported between declawed and control ostriches (Glatz 2010) in inactive, ingestive, grooming, aggression and locomotory behaviour (Table 6.4) of 1-year-old ostriches observed in paddocks. Likewise in another study, no differences in aggressive interactions (Meyer et al. 2003) between declawed birds and control birds in year-old birds were reported, except for chasing, which was more prevalent in the control group.

The concern in ostriches, however, was an increase in slipping (Glatz 2010) of declawed ostriches (Table 6.4). The observations were made on an ostrich farm with 63 cm annual rainfall. The paddocks were wet with considerable forage, which may have caused the declawed ostriches to slip more than control birds when walking around the paddock. Clearly the removal of the toes has a significant impact on the ability of the bird to maintain its footing particularly in wet areas. The concerns noted for behaviour of declawed ostriches in the first 2 weeks after declawing were not observed in the 1-year-old birds.

Several authors have reported greater inactivity in domestic poultry subject to partial beak amputation possibly as a consequence of chronic pain (Duncan and Petherick 1989; Lee and Craig 1990). In year old ostriches neuromas were not observed in stumps of declawed ostriches (Meyer et al. 2003) indicating that year old ostriches are unlikely to be suffering persistent chronic pain. The behavioural evidence for ostriches suggests that declawing does not compromise the locomotor ability of ostriches except in wet regions (Glatz 2010) and has the benefit of improving skin quality, by reducing scratch and kick marks (Meyer et al. 2002, 2003). Human amputees report not only phantom pain and stump pain, but also report phantom sensations, which give the feeling that the amputated limb is still present (Jensen et al. 1983, 1984). Therefore, it is still possible that 1-year-old ostriches may experience some phantom pain sensations in the toe stump.

6.4.2.4 Behaviour of 1-Year-Old Declawed Emus

A study examined whether declawing of emus at day old affected the subsequent locomotor and general behaviour of 1-year-old emus compared to control birds not declawed (Glatz 2001b). Inactive, ingestive, posture change, grooming, aggressive and locomotors behaviour were monitored on year-old birds over a 3-week period during summer in a semi-arid region of South Australia.

There was no behavioural evidence to indicate loss of locomotor ability of declawed emus. The declawed emus engaged (Fig. 6.2) in significantly more

Table 6.4 Effect of declawing on bouts and time (seconds) year old ostriches engaged in inactive, ingestive, grooming, aggressive and locomotor behaviours over 30 min

Variable	Declaw	Control	*P*	LSD
Inactive				
Sit down bouts	0.55	0.56	0.98	NS
Sit down time	354	350	0.98	NS
Sit up bouts	0.05	0.06	0.94	NS
Sit up time	33	28	0.97	NS
Stand bouts	0.20	0.28	0.63	NS
Stand time	128	137	0.61	NS
Ingestive				
Forage bouts	11.9	14.7	0.36	NS
Forage time	606	508	0.53	NS
Eat bouts	5.05	7.27	0.27	NS
Eat time	247	322	0.55	NS
Drink bouts	0.30	0.56	0.49	NS
Drink time	4.45	4.44	0.99	NS
Eliminate	0.30a	0.89b	0.006	0.4
Grooming and other behaviours				
Change posture	0	0	–	–
Preen bouts	3.0	2.83	0.89	NS
Preen time	43	27	0.42	NS
Preen others	0	0.06	0.29	NS
Head scratch	0.10	0.17	0.62	NS
Head shake	0.60	0.35	0.41	NS
Exhibition bouts	0.10	0.29	0.20	NS
Head through fence	0.10	0.64	0.32	NS
Peck environment bouts	1.90	2.65	0.46	NS
Peck environment time	6.9	9.0	0.58	NS
Aggressive behaviours				
Run chase bouts	0	0.11	0.13	NS
Run chase time	0	0.7	0.17	NS
Run away bouts	0.15	0.33	0.25	NS
Runaway time	1.75	2.8	0.58	NS
Locomotor				
Walk bouts	8.95	12.11	0.09	NS
Walk time	267	321	0.44	NS
Pace bouts	0.10	0	0.35	NS
Pace time	0.39	0.40	0.97	NS
Run bouts	0.39	0.40	0.97	NS
Run time	6	2.8	0.51	NS
Bump	0.05	0	0.35	NS
Slip	0.25a	0b	0.022	0.2

Means within rows followed by the same letter are not significantly different at $P = 0.05$; *LSD* least significant difference, *P* probability value from analysis of variance

bouts and time of searching and less stereotyped pacing than control birds (Fig. 6.3) suggesting they were under less stress or less frustrated. Control birds (not declawed) engaged in more pushing of other birds and higher levels of stereotyped behaviour (pecking at fence posts, pacing) presumably as a method to cope in an environment where other birds were seeking dominance (Glatz 2001b).

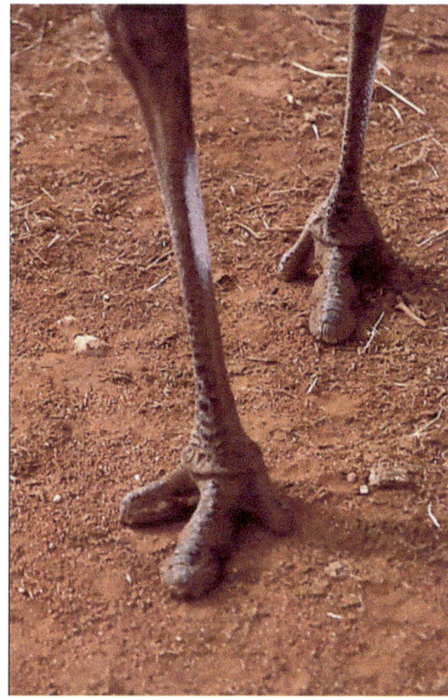

Fig. 6.2 Declawed emu

Fig. 6.3 Emu with intact claws

Pacing is considered to be generally observed in confined and restricted animals in which the animals are often unable to express or perform their natural behavioural pattern freely. Confinement and restrictions, which deny the exercise of natural behavioural needs, could be frustrating (Duncan and Wood-Gush 1972) leading to the re-direction or substitution of behavioural stereotypies (Rushen 1984, 1985; Odberg 1987). It has been suggested that stereotypes could be a positive mechanism to enable the animal to cope with the environment although it is not clear whether the stereotypes themselves are the source of coping (Mason 1991) or a sign of habituation. The greater levels of stereotyped pacing observed in control emus could be the mechanism enabling the bird to cope with the stress of being in a threatening environment where they fear the attack from other emus with intact claws. Emus may have analogous behaviour to poultry and an ostrich (McKeegan and Deeming 1997) as it is well-known that with increased levels of frustration, behavioural changes observed include increased displacement, stereotyped pacing and increased aggression by dominant birds (Duncan and Wood-Gush 1972).

In less painful arthritic conditions it might be expected that birds would change their posture less frequently in an attempt to achieve more comfort. For instance if the emu was in pain it would tend to be restless when both in the standing and sitting position and also make more frequent changes from the standing to the sitting position and vice versa. No evidence could be found that the declawed emus engaged in fewer posture changes as a result of feeling discomfort compared to the control emus (Glatz 2001b).

Control emus gave more threats by stepping towards other emus and pushing them away and showed more frustration by engaging in more stereotyped behaviours (pacing and head through fence behaviour), which often leads to increased aggression particularly by dominant birds. Declawed emus were no different in their ingestive and grooming behaviours providing further evidence that the practice of declawing did not have a major long term influence on emu behaviour (Glatz 2001b).

6.4.2.5 Skin Quality of Declawed Ostriches

The skin quality of declawed ostriches versus control birds was assessed on a commercial ostrich farm (Glatz 2010). About 400 birds were declawed on the day of hatch by removing the distal phalangeal joint using a Lyon beak-trimming machine; another group of 300 ostriches were not declawed. Birds were transported at 14 months-of-age to an abattoir and graded according to industry standards. Skin quality (grading system of one (good) to four (poor)) of declawed birds was compared to a control group of birds.

A defect in a hide can be a hole, a scratch, a loose scab, a healed wound or bacterial damage. The crown is the area with quills, except in the neck, down to wing fold and also the stomach quill area. For grading purposes the crown is divided into four quarters. The lines dividing the crown area into the four quarters are

25 mm wide. The vertical line stretches from the base of neck to the bottom of crown and the horizontal line stretches between the widest quills on either side of crown area. The grading system used was as follows:

First Grade

A defect in one of the quarters as long as it is not larger than approximately 40×40 mm. At least three quarters must be free from defects. Defects on the cutting lines do not affect the grade. A few less visible scars are allowed as long as they are outside the crown area.

Second Grade

A skin with defects affecting two quarters. At least half of the skin must be free from defects. Visible defects outside the crown area are allowed and will not affect the grading.

Third Grade

Atleast one quarter of the skin must be free from defects. Visible defects outside the crown area are allowed.

Fourth Grade

At least one quarter of the skin must be free from defects. Extensive visible defects outside the crown area are allowed and will not affect grading.

The average skin grade achieved was significantly better (average score of 1.59) for the declawed birds compared to the control birds (average score of 1.76). In particular there was a 12% increase in grade one skins observed for the declawed birds.

6.4.2.6 Skin Quality of Declawed Emus

Declawing emu chicks at day-old delays the time the chicks take to commence feeding and reduces live weight by about 10% at 2 weeks-of-age, but body weight recovers by 3 weeks of age (O'Malley and Snowden 1999). Lesions from emu claws are a major cause of skin damage resulting in a reduction in value of the skin (O'Malley and Snowden 1999). Declawing emu chicks immediately after hatch using a hot blade method reduces skin damage. The birds are easier to handle and risk of injury to farmers is greatly reduced (O'Malley and Snowden 1999).

6.4.3 Feather Pecking in Emus

Feather pecking in emus has not been studied as extensively as in domestic poultry in which it is considered as an abnormal behaviour resulting in extensive damage to the plumage of birds (Hughes and Michie 1982). It is usually a problem associated with confinement of animals, which contrasts with the bird's original habitat. The confinement of emus to paddocks can lead to intensive social interaction and more agonistic acts (Glatz 2001b). Feather pecking is generally accepted as a misdirected behaviour, due to the lack of environmental stimuli, although there is disagreement about its development. Several investigators (Blokhuis 1986; Blokhuis and Arkes 1984; Blokhuis and Van der Haar 1989, 1992; Braastad 1990) have confirmed the relationship with ground pecking and foraging behaviour (Blokhuis 1986). However, Vestergaard and Limburg (1993) associate it with dust bathing behaviour and thus the provision and experience of attractive stimuli, like sand and peat could reduce feather pecking.

Emus, which peck other emus also engaged in more bouts and time foraging (Glatz 2001b). The confinement of emus in small paddocks and the lack of reward for them while attempting to forage in bare paddocks may have lead to frustration resulting in the aggressive pecking behaviour. Birds spend most of their time in beak related activities (Hughes and Grigor 1996) and when there is a very limited activity choice there is more time for feather pecking. Emus being pecked were also found to have a higher incidence of stereotyped behaviours such as fence pecking and head shaking. It is believed the reduction of fear is reflected in adaptive or displacement behaviour.

In the emu study (Glatz 2001b), head shaking and fence pecking were the significant stereotyped behaviours related to birds that were being pecked. Head shaking has been characterised as a coping mechanism and as a symptom of being "better off" (Duncan 1970; Mauldin and Siegel 1979) and is a prominent behaviour in White Leghorn birds (Webster and Hurnik 1990). Head shaking also coincided with the exhibition behaviour in emus, which in itself could attract aggressive pecking by other birds (Glatz 2001b). Emus under threat also spent more time pacing, which is a stereotyped behaviour indicating frustration. In a similar manner, emu giving thrusts were also those birds pecking and chasing other birds but also receiving thrusts and running away (Glatz 2001b).

6.4.4 The Importance of the Stockpersons in Ratite Welfare

Methods to improve skin quality in emus and ostriches involves training birds to enter portable pens by themselves rather than being herded by stockpersons into the pen causing stress to the bird (Glatz 2001a). Likewise, recommendations were provided on moving birds around the farm by training birds to follow the animal handler rather than loading them into transport crates and risking stress and injury to the birds from loading, unloading and transport. While there has been no research undertaken in the ratite industry on skills of stockpersons it is essential that animal handlers have the skills to look after and handle their stock (Hemsworth and Coleman 1998). An issue that has a significant influence on welfare is the type of relationship that exists between the stockperson and their animals. If the stockperson has a good affinity with his animals the welfare and productivity of animals can be improved. This has been demonstrated in poultry, pigs, dairy cows and calves (Hemsworth et al. 1989; Breuer et al. 2000; Lensink et al. 2001; Hemsworth 2003).

Therefore an increasing emphasis by ratite stockpersons on handling birds with care and developing a good relationship with the birds is likely to improve their welfare. It has been demonstrated in other species that poor interactions by stock handlers with livestock will result in animals becoming fearful of handlers (Hemsworth and Coleman 1998) and lead to both acute and chronic stress (Moberg 2000; Hemsworth and Coleman 1998). The approach used by the stockperson when tending to livestock has an influence on the type of response shown by animals. Gentle handling of the animal will improve the relationship between the animal and the handler, whereas rough handling will cause the animal to be wary of the handler (Waiblinger et al. 2006). The approach adopted by handlers with animals largely depends on the attitude they have towards various aspects of their job (Fishbein and Ajzen 1975; Ajzen 1988), which has been demonstrated in a number of livestock species (Waiblinger et al. 2002; Lensink et al. 2000; Hemsworth et al. 1989, 2000, 2002; Coleman et al. 2000, 2003; Breuer et al. 2000).

If stockpersons believe that yelling at and forcing animals is the best way to move animals into a yard, these stockpersons are more likely to handle the animals in a rough and aggressive manner. On the other hand, a stockperson who believes a gentle and calm approach is needed to herd animals is likely to have a better affinity with the animals. A high proportion of poor interactions between the stockperson and the livestock will result in the animals showing fear. The animals are likely to become stressed, which could result in reduced productivity in farm animals. Therefore the attitudes of the stockperson can influence the welfare and productivity of farm animals.

A concern in the ratite industry is that birds are not conditioned to handling (Glatz 2001a) resulting in fractious behaviour of birds and damage to skins. Likewise, there was a concern that chicks were not achieving their early growth potential indicating a need to establish methods to stimulate chicks to eat by using more regular handling, maintaining an optimum environment and socialising the birds to human presence. This requires the handler to visit the chick area often to stir the feed and train the birds to respond positively to human presence. In particular, as the birds' age they should be regularly provided feed and water in portable yards. This encourages birds to pen themselves, which is better than birds being forced into yards and risking injuries to the skin. Moreover, there is a need to reduce the level of aggression in emus and ostriches by providing an enriched environment. This requires the provision of shaded sandy areas for the birds to dust bathe, the use of novel objects in the environment to reduce boredom and the development of stereotype behaviours and aggressive pecking and kicking (Glatz 2001a).

6.4.5 *Housing*

One of the major welfare issues for ratites is the poor air quality in housing when young ratites are being brooded (Glatz 2001a). In particular at night under cold outdoor conditions the houses are usually sealed to prevent loss of heat. This has resulted in birds being exposed to high ammonia levels from wet droppings, contributing to hock burn and breast blisters. On the other hand under dry conditions, dust is a problem in the shed and can reduce chick growth by depressing the immune system. Respiratory infections that can result in lung lesions are increased by contaminants in the air. Sawdust can result in high dust levels and health problems. Poor ventilation in enclosed buildings results in accumulation of ammonia, which can influence the health of chicks; 10–15 ppm of ammonia level in the air of the shed can be smelled and long term exposure to 25–36 ppm of ammonia level will cause eye and nasal irritation in humans (Glatz 2001a). In domestic poultry aviary housing systems have higher ammonia emissions and higher dust levels (Barnett 1999) than conventional cage systems.

The main airborne pollutants found in livestock buildings (including ratite brooding areas) are ammonia, carbon dioxide, dust, other particles and micro-organisms (Wathes et al. 1998). The particles in a livestock house consist of skin, feathers, faeces, bedding, micro-organisms, fungal spores, pollen, feed, and other small particles from outdoors (Takai et al. 1998). The airborne particles can carry bacteria, viruses, endotoxins, odorous materials and gases (Seedorf et al. 1998). There is an increasing concern about the effects of respirable particles on human and animal health. Dust, bedding, faeces and feed is a major concern in poultry buildings (Van Wicklen et al. 2001; Mitchell et al. 2002; Ikeguchi 2002; Kristensen et al. 2000; Wathes et al. 1998) contributing to most of the inhalable and respirable particles and may also be a concern in ratite housing. There is a need for a comprehensive examination of air quality in ratite housing facilities to determine the impact on bird health and growth particularly during the chick brooding stage.

6.4.5.1 Impact of Housing on Animal Health, Production and Welfare

In domestic poultry there is a strong relationship between production and welfare (Al Homidan et al. 1998; Quarles and Caveny 1979; Feddes et al. 1995; Hayter and Besch 1974; Kristensen et al. 2000). The immunological challenges often associated with poor air quality can lead to a reduction in feed intake and production (Kelley et al. 1987; Kemeny 2000). Airborne particles could also increase the susceptibility of birds to diseases by their irritant action or via allergic reactions (Harry 1978). It is likely that improving air quality in ratite brooding facilities could improve production and provide a better working environment for stockpersons.

6.5 Conclusions

There is an urgent need to examine alternatives to declawing in ratites while conducting a thorough welfare evaluation of pain and fear responses in declawed birds. In addition, other methods of declawing need to be studied, especially the effectiveness of a microwave method of declawing on ostrich behaviour and skin quality. There is considerable development of knowledge of the factors affecting internal pollutant concentrations in poultry buildings, but little data exist in the ratite industries. Studies are required to identify the factors that have the greatest influence on pollutants in ratite housing facilities and methods to reduce the high levels of ammonia. It is important that there is an investigation of the relationship that exists between ratite stockpersons and their birds and to measure the responses of ratites to humans and the impact on the birds fear, growth and skin quality.

References

Ajzen I (1988) Attitudes, personality and behaviour. Open University Press, Chicago

Al Homidan A, Robertson JF, Petchey AM (1998) Effect of environmental factors on ammonia and dust production and broiler performance. Br Poult Sci 39 (Supplement):S10

Animal Welfare Advisory Committee (1998) Code of recommendations and minimum standards for the welfare of ostrich and emu. http://www.biosecurity.govt.nz/animal-welfare/codes/ostriches-emus/index.htm#1. Cited 18 June 2010

Anon (1992) Report on priorities for animal welfare research and development. Farm Animal Welfare Council, Ministry of Agriculture, Fisheries and Food, Surbiton, UK

Barnett J (1999) Evaluation of alternative egg laying production systems in Europe (August/September 1999) A Travel Report Presented to the Rural Industries Research and Development Corporation, Canberra, Australia

Barnett JL, Newman EA (1997) Review of the welfare research in the laying hen and the research and management implications for the Australian egg industry. Aust J Agric Res 48:385–402

Barnett JL, Glatz PC, Almond A, Hemsworth PH, Cransberg PH, Parkinson GB, Jongman EC (2001) A Welfare Audit for the Chicken Meat Industry. Report to the Rural Industries Research and Development Corporation, Canberra, Australia

Blokhuis JH (1986) Feather pecking in poultry: its relation to ground pecking. Appl Anim Behav Sci 16:63–67

Blokhuis JH, Arkes JG (1984) Some observations on the development of feather pecking in poultry. Appl Anim Behav Sci 12:145–157

Blokhuis JH, Van Der Haar WJ (1989) Effects of floor type during rearing and of beak trimming on ground pecking and feather pecking in laying hens. Appl Anim Behav Sci 22:359–369

Blokhuis JH, Van Der Haar WJ (1992) Effects of pecking incentives during rearing on feather pecking of laying hens. Br Poult Sci 33:17–24

Braastad BO (1990) Effects on behaviour and plumage of a key stimuli floor and a perch in triple cages for laying hens. Appl Anim Behav Sci 27:127–139

Brambell FWR, Barbour DS, Barnett MB, Ewer TK, Hobson A, Pitchforth H, Smith WR, Thorpe WH, Winship FJW (1965) Report of the Technical Committee to Enquire into the Welfare of Animals Kept Under Intensive Husbandry Systems. Her Majesty's Stationery Office, London

Breuer K, Hemsworth PH, Barnett JL, Matthews LR, Coleman GJ (2000) Behavioural response to humans and the productivity of commercial dairy cows. Appl Anim Behav Sci 66:273–288

Breward J, Gentle MJ (1985) Neuroma formation and abnormal afferent nerve discharges after partial beak amputation (beak-trimming) in poultry. Experientia 41:1132–1135
Broom DM (1986) Indicators of poor welfare. Br Vet J 142:524–526
Christensen JW, Nielsen BL (2004) Environmental enrichment for ostrich, *Sruthio camelus*, chicks. Anim Welf 13(2):119–124
Coleman GJ, Hemsworth PH, Hay M, Cox M (2000) Modifying stockpersons attitudes and behaviour towards pigs at a large commercial scale farm. Appl Anim Behav Sci 66:11–20
Coleman GJ, McGregor M, Hemsworth PH, Boyce J, Dowling S (2003) The relationship beliefs, attitudes and observed behaviours of abattoir personnel in the pig industry. Appl Anim Behav Sci 82:189–200
Communities E (1998) Council Directive 98/58/EC of 20 July 1998 concerning the protection of animals kept for farming purposes. Off J L 221:23–27
Dawkins MS (1980) Animal suffering: the science of animal welfare. Chapman and Hall, London
Dawkins MS (1983) Battery hens name their price: consumer demand theory and the measurement of animal needs. Anim Behav 31:1195–1205
Dawkins MS (2003) Behaviour as a tool in the assessment of animal welfare. Zoology 106:383–387
Duncan IJH (1970) Frustration in the fowl. In: Freeman BM, Gordon RF (eds) Aspects of poultry behaviour. Oliver and Boyd, Edinburgh, pp 15–31
Duncan IJH, Fraser D (1997) Understanding animal welfare. In: Appleby MC, Hughes BO (eds) Animal welfare. CAB International, Oxon, pp 19–31
Duncan IJH, Petherick JC (1989) Cognition: the implications for animal welfare. Appl Anim Behav Sci 24:81
Duncan IJH, Wood-Gush DGM (1972) The analysis of displacement preening in the domestic fowl. Anim Behav 20:68–71
Feddes JJR, Taschuk K, Robinson FE, Riddell C (1995) Effect of litter oiling and ventilation rate on air quality, health and performance of turkeys. Can Agr Eng 37(1):57–62
Fishbein M, Ajzen I (1975) Belief, attitude, intention and behaviour. Addison-Wesley Publishing Company, Reading, Massachusetts
Food and Agriculture Organization of the United Nations (2009) Capacity building to implement good animal welfare practices. Report of the FAO Expert Meeting, FAO Headquarters (Rome). ftp://ftp.fao.org/docrep/fao/012/i0483e/i0483e00.pdf
Fraser D (2003) Assessing animal welfare at the farm and group level: the interplay of science and values. Anim Welf 12:433–443
Gentle MJ (1986) Neuroma formation following partial beak amputation (beak trimming) in the chicken. Res Vet Sci 41:383–385
Glatz PC (2001a) Improving skin quality of emus and ostriches. A benchmark study of husbandry, transport, lairage and slaughter methods. Rural Industries Research and Development Corporation Publication No 01/04, ISBN 0 642 58227 0, ISSN 1440–6845
Glatz PC (2001b) Effect of declawing on behaviour of farmed emus. Asian-Aust J Anim Sci 14:288–296
Glatz PC (2008) Ratite toe-trimming-for ostriches and emus-Training manual. Rural Industries Research and Development Corporation Publication No 08/017, ISBN 1 74151 606 4, ISSN 1440–6845
Glatz PC (2010) Husbandry and genetic strategies to improve hide quality of ostriches. Rural Research and Development Corporation Publication No 09/174, ISBN 1 74151971 3, ISSN 1440–6845
Glatz PC, Miao ZH (2008) Husbandry of ratites and potential welfare issues: a review. Aust J Exp Agr 48(10):1257–1265
Grigor PN, Hughes BO, Gentle MJ (1995) Should turkeys be trimmed? An analysis of the costs and benefits of different methods. Vet Rec 136:257–265
Harry EG (1978) Air pollution in farm building and methods of control: a review. Avi Pathol 7:441–454
Hayter RB, Besch EL (1974) Airborne-particle deposition in the respiratory tract of chickens. Poult Sci 53:1507–1511

Hemsworth PH (2003) Human-animal interactions in livestock production. Appl Anim Behav Sci 81:185–198
Hemsworth PH, Coleman GJ (1998) Human–livestock interactions. The stockperson and the productivity and welfare of intensively farmed animals. CAB International, Wallingford, UK
Hemsworth PH, Barnett JL, Matthews LR, Coleman GJ, Hansen C (1989) A study of the relationship between the attitudinal and behavioural profiles of stockpersons and the level of fear of humans and reproductive performance of commercial pigs. Appl Anim Behav Sci 23:301–314
Hemsworth PH, Coleman GJ, Barnett JL, Borg S (2000) Relationships between human-animal interactions and productivity of commercial dairy cows. J Anim Sci 78:2821–2831
Hemsworth PH, Coleman GJ, Barnett JL, Borg S, Dowling S (2002) The effects of cognitive behavioural intervention on the attitude and behaviour of stockpersons and the behaviour and productivity of commercial dairy cows. J Anim Sci 80:68–78
Hughes BO, Grigor PN (1996) Behavioural time budgets and beak related behavior in floor housed turkeys. Anim Welf 5(2):189–198
Hughes BO, Michie W (1982) Plumage loss in medium bodied hybrid hens: The effect of beak trimming and cage design. Poult Sci 23:59–64
Ikeguchi A (2002) Ultra sonic sprayer controlling dust in experimental poultry houses. Agr Eng Int, Manuscript BC 01 002, Vol IV
Jensen TS, Krebs B, Nielsen J, Rasmussen P (1983) Phantom limb, phantom pain and stump pain in amputees during the first 6 months following limb amputation. Pain 17:243–256
Jensen TS, Krebs B, Nielsen J, Rasmussen P (1984) Non-painful phantom limb phenomena in amputees: incidence, clinical characteristics and temporal course. Acta Neurol Scand 70:407–414
Kelley KW, Brief S, Westly HJ, Novakofski J, Bechtel PJ, Simon J, Walker ER (1987) Hormonal regulation of the age-associated decline in immune function. Ann NY Acad Sci 496:91–97
Kemeny DM (2000) The effects of pollutants on the allergic immune response. Toxicology 152(1–3): 3–12
Kristensen HH, Burgess LR, Demmers TGH, Wathes CM (2000) The preferences of laying hens for different concentrations of atmospheric ammonia. Appl Anim Behav Sci 68:307–318
LayWel (2006) Welfare implication of changes in production systems for poultry. http://www.laywel.eu/web/ date accessed
Lee HY, Craig JV (1990) Beak-trimming effects on the behaviour and weight gain of floor-reared, egg strain pullets from three genetic stocks during the rearing period. Poult Sci 69:568–575
Lensink BJ, Boissy A, Veissier I (2000) The relationship between farmers' attitude and behaviour towards calves, and productivity of veal units. Ann Zootech 49:313–327
Lensink BJ, Fernandez X, Cozzi G, Florand L, Veissier I (2001) The influence of farmers' behaviour on calves' reactions to transport and quality of veal meat. J Anim Sci 79:642–652
Lunam CA, Glatz PC (2000) Declawing of farmed emus. Harmful or Helpful. A final report to the Rural Industries Research and Development Corporation. Rural Industries Research and Development Corporation Publication No 99/177, ISBN 0 642 58026 X ISSN 1440–6845
Mason GJ (1991) Stereotypies: a critical review. Anim Behav 41:1015–1037
Mauldin JM, Siegel PB (1979) Fear, head shaking and production in five populations of caged chickens. Br Poult Sci 20:39–44
McKeegan DEF, Deeming DC (1997) Effects of gender and group size on the time-activity budgets of adult breeding ostriches (*Struthio camelus*) in a farm environment. Appl Anim Behav Sci 51:159–177
Mellor DJ, Cook CJ, Stafford KJ (2000) Quantifying some responses to pain as a stressor. In: Moberg GP, Mench JA (eds) The biology of animal stress. CAB International, Oxon, pp 171–198
Meyer A, Cloete SWP, Brown CR, van Schalkwyk SJ (2002) Declawing ostrich (*Struthio camelus domesticus*) chicks to minimize skin damage during rearing. S Afr J Anim Sci 32:192–200
Meyer A, Cloete SWP, Brown CR, van Schalkwyk SJ (2003) The persistence to slaughter age of scars resulting from damage inflicted to ostrich skins during the grow-out phase. S Afr J Anim Sci 33:32–37

Mitchell BW, Buhr RJ, Berrang ME, Bailey JS, Cox NA (2002) Reducing airborne pathogens, dust and *salmonella* transmission in experimental hatching cabinets using an electrostatic space charge system. Poult Sci 81:49–55

Moberg GP (2000) Biological response to stress: implications for animal welfare. In: Moberg GP, Mench JA (eds) The biology of animal stress. CABI Publishing, CAB International, Oxon and New York, pp 1–21

O'Malley P, Snowden I (1999) Emu products. Increasing production and profitability. A report for the Rural Industries Research and Development Corporation. Research Paper Series No 99/143, Canberra, Australia

Odberg F (1987) Behavioural responses of stress in farm animals. In: Van Adrichem PWM, Wiepkema PR (eds) The biology of stress in farm animals: an integrated approach. Martinus Nijhoff, Dordrecht, pp 135–149

Office of the Queensland Parliamentary Counsel (2009) *Animal Care and Protection Act 2001*. Animal Care and Protection. Regulation 2002. Reprint No 3B. http://www.legislation.qld.gov.au/LEGISLTN/CURRENT/A/AnimalCaPrR02.pdf

Quarles CL, Caveny DD (1979) Effects of air contaminants on performance and quality of broilers. Poult Sci 58:543–548

Rushen J (1984) Stereotyped behaviour, adjunctive drinking and the feeding periods of tethered sows. Anim Behav 52:1059–1067

Rushen J (1985) Stereotypies, aggression and the feeding schedules of tethered sows. Appl Anim Behav Sci 14:137–147

Sandøe P, Forkman F, Christiansen SB (2004) Scientific uncertainty - how should it be handled in relation to scientific advice regarding animal welfare issues? Anim Welfare 13:121–126

Seedorf J, Hartung J, Schroder M, Linkert KH, Phillips VR, Holden MR, Sneath RW, Short JL, White RP, Pedersen S, Takai H, Johnsen JO, Metz JHM, Groot Koerkamp PWG, Uenk GH, Wathes CM (1998) Concentrations and emissions of airborne endotoxins and microorganisms in livestock buildings in Northern Europe. J Agric Eng Res 70:97–109

Standing Committee on Agriculture and Resource Management (2003) Model code of practice for the welfare of animals. Farming of ostriches. CSIRO Publications, East Melbourne, Australia

Takai H, Pedersen S, Johnsen JO, Metz JHM, Groot Koerkamp PWG, Uenk GH, Phillips VR, Holden MR, Sneath RW, Short JL, White RP, Hartung J, Seedorf J, Schroder M, Linkert KH, Wathes CM (1998) Concentrations and emissions of airborne dust in livestock buildings in Northern Europe. J Agric Eng Res 70:59–77

Uhart M, Aprile G, Beldomenico P, Solis G, Marull C, Beade M, Carminati A, Moreno D (2006) Evaluation of the health of free-ranging greater rheas (*Rhea americana*) in Argentina. Vet Rec 158:297–303

Van Wicklen GL, Foutz TL, Rowland GN (2001) Respirable tissue damage in broilers exposed to aerosol particles and ammonia. Trans ASAE 44:889–1894

Vestergaard KS, Limburg L (1993) A model of feather pecking development which relates dust bathing in the fowl. Behaviour 26:291–308

Waiblinger S, Menke C, Coleman GJ (2002) The relationship between attitudes, personal characteristics and behaviour of stockpeople and subsequent behaviour and production of dairy cows. Appl Anim Behav Sci 79:195–219

Waiblinger S, Boivin X, Pedersen V, Tosi MV, Janczak AM, Visser EK, Jones RB (2006) Assessing the human-animal relationship in farmed species: a critical review. Appl Anim Behav Sci 101:185–242

Wathes CM, Phillips VR, Holden MR, Sneath RW, Short JL, White RPP, Hartung J, Seedorf J, Schroder M, Linkert KH (1998) Emissions of aerial pollutants in livestock buildings in northern Europe: overview of a multinational project. J Agric Eng Res 70:3–9

Webster AB, Hurnik JF (1990) Behaviour, production, and well being of the laying hen. 2. Individual variation and relationships of behaviour to production and physical condition. Poult Sci 70:421–428

Weeks CA, Nicol CJ (2006) Preferences of laying hens. Worlds Poult Sci J 62:296–307

Chapter 7
The Structure and Sensory Innervation of the Integument of Ratites

K.A. Weir and C.A. Lunam

Abstract This chapter reviews the microanatomy and innervation of the integument of the ostrich and the emu. We consider how these structures enable ratites to interact with their environment and how damage to the skin may compromise the welfare of the birds.

The skins of the ostrich and the emu are structurally very similar. The epidermis is very thin and therefore provides little protection against mechanical injury. Waterproofing of the skin is aided by keratin and lipids produced by sebokeratinocytes.

The dense connective tissue of the dermis is organised into two distinct layers: a thin *stratum superficiale* and an extensive *stratum compactum*. The paucity of elastic fibres suggests that both the strength and flexibility of the skin is due to the three-dimensional cross-weave arrangement of the collagen bundles. The high vascularity near the surface explains why the skin bruises easily during normal behaviours, as well as during handling and transport.

Free nerve endings, with peptide combinations similar to mammalian nociceptors, have been reported in emu skin. Injury to the skin is therefore likely to inflict pain via stimulation of these nerves. Herbst corpuscles are mechanoreceptors found near the contour feathers of both the ostrich and emu. Their likely function in ratites is to detect feather movement.

Filoplumes and bristle hairs invariably form a semicircular pattern caudal to the contour feather follicle in ostrich skin. Their interfollicular distribution, however, is highly variable between different flocks. Their potential function as mechanoreceptors, the possibility of a genetic heritability of these traits, and their effects on skin quality after tanning are discussed.

K.A. Weir
School of Biomedical Sciences, University of Queensland, St Lucia, QLD 4072, Australia

C.A. Lunam (✉)
Flinders University, GPO Box 2100 Adelaide, SA 5001, Australia
e-mail: Chris.lunam@flinders.edu.au

P.C. Glatz et al. (eds.), *The Welfare of Farmed Ratites*, Animal Welfare 11,
DOI 10.1007/978-3-642-19297-5_7, © Springer-Verlag Berlin Heidelberg 2011

Keywords Filoplumes · Innervation · Physiology · Ratite integument · Structure · Welfare

7.1 Introduction

Current knowledge of the structure and function of ratite skin is derived from studies of the emu and ostrich with little or no information available regarding the skin of the rhea, cassowary, or kiwi. This disparity in our understanding of skin structure in each of the extant ratite groups is likely due to interest in the ostrich and emu for the production of tanned skins. Similarly, current knowledge of ostrich skin is largely restricted to those layers that are removed at slaughter and subsequently processed for leather.

In the following discussion, we review the current knowledge of the structure and innervation of the ratite integument. Included in our discussion is the specific arrangement of the different tissue types, a specialised sensory receptor, the Herbst corpuscle, and the modified feathers, the filoplumes and bristle hairs. The implications of these structures and their innervation on the physiology and welfare of ratites are considered.

7.2 Structure and Function of the Integument

7.2.1 Epidermis

The most superficial layer of the skin of ratites and other vertebrates is a stratified squamous epithelium termed the epidermis. The epidermis is an avascular, cellular layer that is responsible for a number of crucial body functions including protection of the underlying tissues from mechanical damage (aided by feathers), water loss, chemicals, toxins, parasites, and microbiological attack. All of these functions have an important influence on the welfare of the individual bird. Control of cutaneous water loss may be particularly important for thermoregulation in ratites such as the ostrich and emu. These birds are found in arid habitats, and like other birds, they lack the sweat glands that facilitate evaporative cooling in mammals (Menon et al. 1996; Menon and Menon 2000).

While Frapple et al. (1997) describe the epidermis of the emu as "dense and thick", our studies have revealed only three to five layers of cells that are organised into two strata: the *stratum basale* and *stratum intermedium* (Weir and Lunam 2004). These living cell layers are covered by a keratinised dead cell layer, the *stratum corneum* (Fig. 7.1a). Reports of the numbers of living cell layers in the epidermis of the ostrich vary from two to three in belly skin (Lunam and Weir 2006) to three to seven layers in neck skin (Menon et al. 1996). While the thickness of the

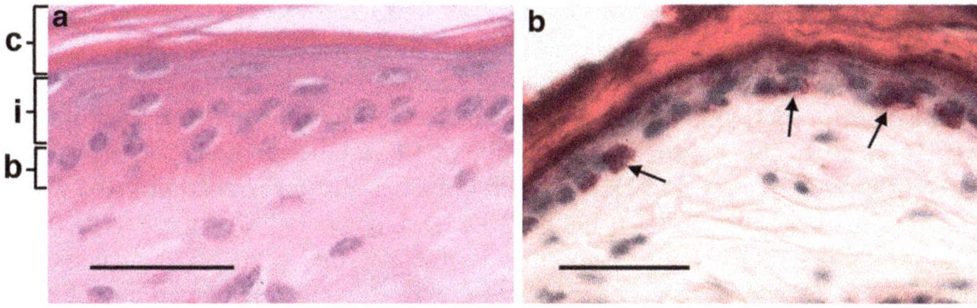

Fig. 7.1 Light micrographs of the epidermis of emu skin. (**a**) Haematoxylin and eosin stained section showing c, *stratum corneum*; i, *stratum intermedium*; and b, *stratum basale*. (**b**) *Oil red* O stained section showing lipid spheres (*arrows*) densely packed within the cells of the *stratum basale*. *Scale bars* = 40 μm

combined *stratum basale* and *stratum intermedium* is consistent across body regions in the emu, the *stratum corneum* of skin adjacent to the wing is thinner than that on the rump and back (Weir 2004). These results suggest that the structure of the epidermis is similar in the ostrich and the emu, but there are differences with body region. A thinner epidermis may provide less resistance to mechanical damage and subsequent bruising. It remains to be determined whether these minor differences in thickness of epidermal layers correlate with differences in the function of the skin in these body regions.

Sebokeratinocytes are the main epidermal cell type in birds. These cells arise from the basal layer of the epidermis and, as their name suggests, accumulate keratin and lipids in their cytoplasm as they differentiate and move towards the *stratum corneum* (Menon et al. 1986, 1996). In birds and mammals, epidermal lipids are important for preventing water loss across the skin; however, the organisation of lipids within the avian *stratum corneum* is different to that reported in mammals. In mammals, lamellar body secretion results in the formation of lipid bilayers between cell remnants of the *stratum corneum* (Elias and Friend 1975; Aszterbaum et al. 1992; Ross and Pawlina 2006), whereas in birds free lipid droplets fill the space between adjacent sebokeratinocytes (Menon et al. 1986). It has been suggested that birds have a relatively poor barrier to water loss due to the reliance on cutaneous evaporation for control of body temperature along with panting (Menon et al. 1996). Furthermore, it appears that some avian species, such as the zebra finch, are able to reduce the level of cutaneous water loss by altering the organisation of epidermal lipids in response to reduced water availability, a process termed facultative waterproofing (Menon et al. 1986, 1996; Menon and Menon 2000). Ultrastructural investigation of the ostrich epidermis in the unfeathered skin of the neck has revealed lipid morphology consistent with a poor barrier to water loss (Menon et al. 1996); however, measurement of cutaneous water loss in ostrich thigh skin has shown varied results, ranging from less than 2% of total evaporative loss (Schmidt-Nielsen et al. 1969) to 40% (Withers 1983). It is unknown whether the ostrich is capable of facultative cutaneous waterproofing in water-deprived conditions as described in the zebra finch (Menon et al. 1996).

In the emu, we found numerous epidermal lipid spheres that are concentrated in the basal layers and increase in size as the cells move towards the *stratum corneum* (Fig. 7.1b; Weir and Lunam 2004). No data are available on the ultrastructural organisation of epidermal lipids in the emu. Interestingly, cutaneous evaporative water loss contributes approximately 30% of the total evaporative loss in adult emus at 45°C (Maloney and Dawson 1994, 1998), but under conditions of water deprivation, this is reduced to below 10% (Maloney and Dawson 1998). It is unknown which body regions are responsible for this reduction in cutaneous water loss and whether this is due to changes in the organisation of epidermal lipids.

Another factor, which may affect cutaneous water loss, is the age of the bird. In contrast to adult zebra finches, nestlings have a good barrier to epidermal water loss (Menon et al. 1988). Data from ostrich chicks suggest they have similar levels of cutaneous water loss compared to adult ostriches (Withers 1983). While cutaneous water loss values in emus of different ages are lacking, our studies have demonstrated that the cellular epidermis (*stratum basale* and *stratum intermedium*) is thicker in emus at 1 week of age compared to 18 months or 3 years of age (Weir 2004). Similarly, the *stratum corneum* is thicker in 1-week-old emus compared to 18 months but is not significantly different to the *stratum corneum* of 3-year-old emus (Weir 2004). To date, the ultrastructure of the epidermis of emu chicks has not been compared to the adult; hence, it is not known whether the change in thickness of the epidermis is accompanied by alteration in the organisation of epidermal lipids with age.

It is clear that more research is required in order to correlate epidermal morphology with cutaneous water loss in ratites of different ages under differing environmental conditions. This will aid our understanding of homeostasis in these birds so that we can be better informed regarding the effects of water availability and ambient temperature on bird welfare. The data available to date support the suggestion of Glatz and Miao (2008) that birds should be allowed enough space to make postural changes so that body temperature can be regulated with changing ambient temperature. This is because sitting down covers the unfeathered skin of the leg region and conserves heat (Maloney and Dawson 1994). Shelter, water availability, and available space should be considered across all aspects of husbandry in order to optimise the well-being of farmed ratites.

7.2.2 Dermis

The dermis lies immediately beneath the epidermis and consists of four major layers in birds. From superficial to deep these are the *stratum superficiale*, *stratum compactum*, *stratum laxum*, and *lamina elastica* (Fig. 7.2; Lucas and Stettenheim 1972). The *stratum superficiale* (grain layer) and *stratum compactum* (corium layer) form the dense collagenous tissue that is removed for the production of tanned ostrich and emu skins. These layers have been described extensively in both the ostrich and emu. The *stratum laxum* is a thick adipose tissue layer, which holds the base of the feather follicles, the smooth muscles that move the feathers

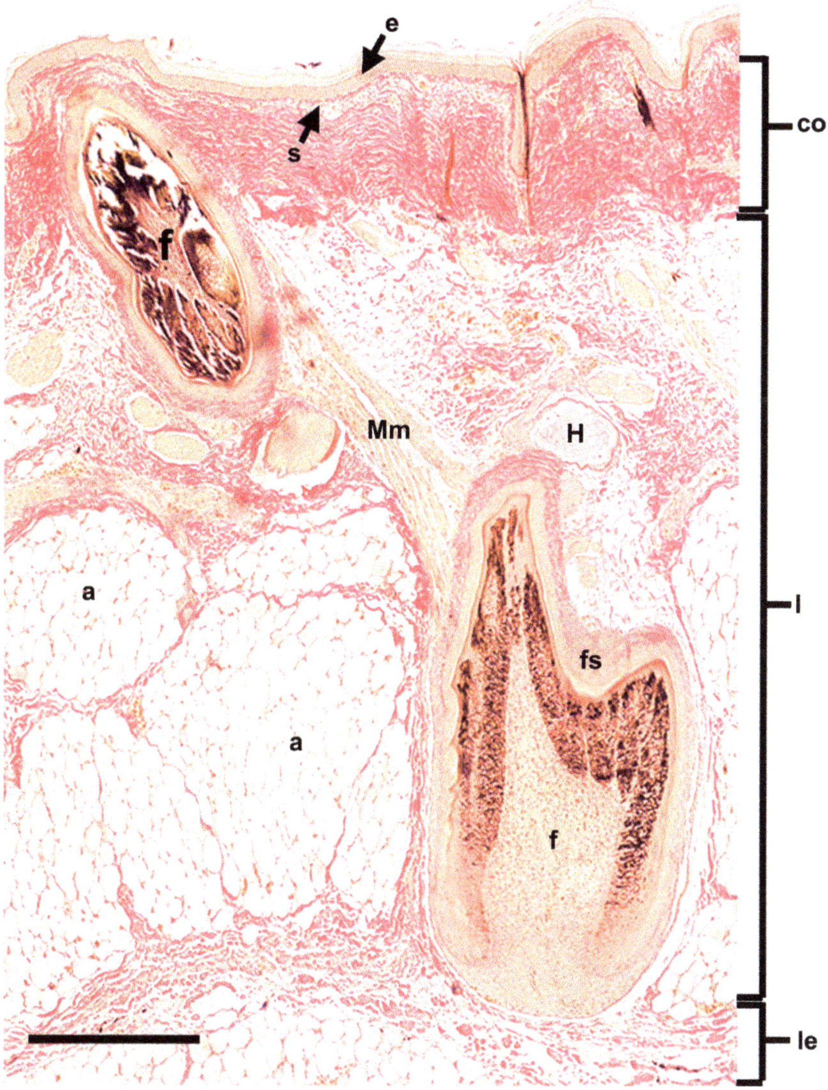

Fig. 7.2 Light micrograph showing a section through emu chick skin. *e* epidermis; *s* stratum superficiale; *co* stratum compactum; *l* stratum laxum; and *le* lamina elastica. Deep to the *lamina elastica* is the subcutis. *The stratum laxum* contains adipose tissue (a) that supports the feather (f) and the follicle sheath (fs) with the attached pennamator muscles (Mm). A Herbst corpuscle (H) is visible adjacent to the sheath. Verhoeff and van Gieson stain. *Scale bar* = 200 μm

(*Mm. pennarum*), as well as the larger branches of blood vessels and nerves that supply the skin. The lamina elastica is a thin, elastic layer that defines the deep border of the integument. The *stratum laxum* and lamina elastica of ratites have been described only in the emu (Weir and Lunam 2004).

The *stratum superficiale* or grain layer is relatively thin in both the ostrich and emu (Figs. 7.2 and 7.3). Frapple et al. (1997) first described the grain layer of the emu as thin, dense, and poorly connected to the underlying dermis. Our light microscopy and scanning electron microscopy studies in both the ostrich (Lunam and Weir 2006) and emu (Weir and Lunam 2004) confirm these observations. Furthermore, in the emu we have shown that the *stratum superficiale* comprises approximately 1% of

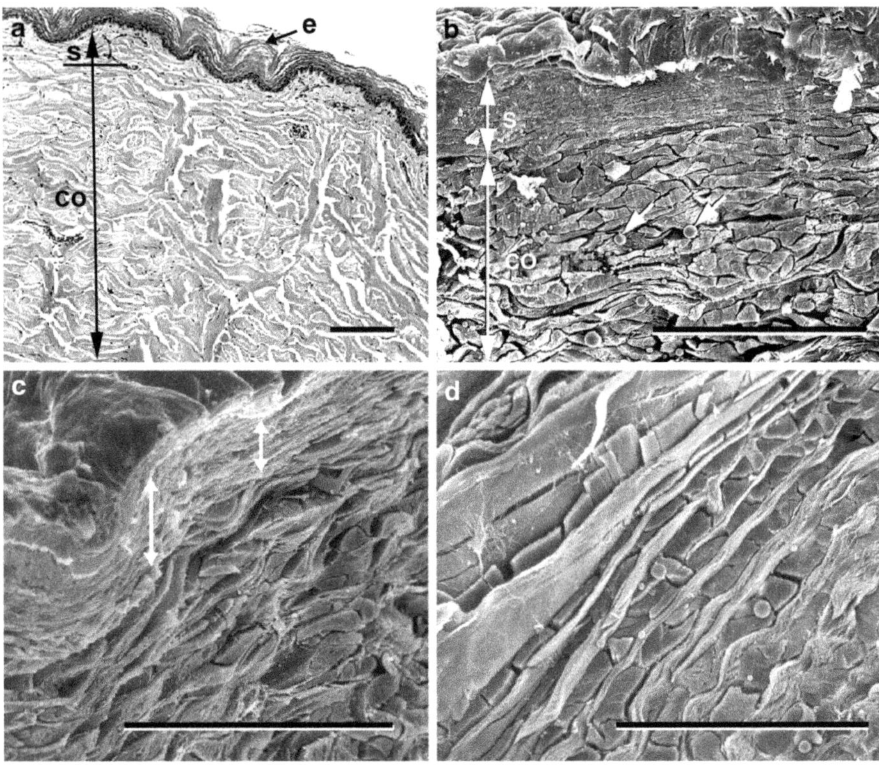

Fig. 7.3 (**a–b**) Ostrich skin; *e* stratum corneum of the epidermis; *s* stratum superficiale; and *co* stratum compactum. (**a**) Light micrograph showing layers of epidermal cells as a dark granular band beneath the *stratum corneum*. Immediately beneath the epidermis lies the dermis consisting predominantly of collagen fibres. Compared to the *stratum compactum*, the *stratum superficiale* is relatively thin and highly vascularised. Red blood cells within blood vessels are seen as clusters of *black dots* throughout the dermis. Haematoxylin and eosin stain. (**b**) Scanning electron micrograph showing the clear differentiation between the densely packed collagen bundles in the *superficiale* layer compared to the larger bundles of the adjacent *compactum*. Fat globules (*arrows*) are present in the spaces between the collagen bundles. (**c–d**) Scanning electron micrographs of emu skin. (**c**) The *stratum superficiale* (*arrows*) consists of a thin layer of collagen bundles orientated predominantly parallel to the surface. (**d**) Cross-section through the dermis showing the highly organised arrangement of collagen bundles running at right angles to each other in consecutive layers. *Scale bars* = 100 μm. (**a**) and (**b**) reproduced with permission from Lunam and Weir (2006) Rural Industries Research and Development Corporation publication No. 06/054

total skin thickness (35 ± 14.3 μm; Weir and Lunam 2004). In both the ostrich (Lunam and Weir 2006) and emu (Weir and Lunam 2004), the collagen fibres of the *stratum superficiale* are much thinner than those in the *stratum compactum* and most of these fibres run parallel to the surface of the skin with very few serving to anchor the two layers together. These factors are likely to contribute to loose grain in tanned skins. The absence of dermal papillae in emu skin (Weir and Lunam 2004) indicates that there is also a relatively loose connection between the dermis and epidermis and suggests that the skin is susceptible to damage from shearing forces. Lange (1929) reported collagen structures at the dermal–epidermal border in the ostrich and these have been interpreted as dermal papillae (Bezuidenhout 1999); however, we have not found any distinct dermal papillae in the ostrich skin (Fig. 7.3a; Lunam and Weir 2006). These differences may be due to the examination of skin from different body regions or due to the authors' interpretation of dermal papillae.

At the border between the *stratum superficiale* and *stratum compactum* is a thin strip of loose connective tissue that is richly supplied with small blood vessels (Frapple et al. 1997; Lunam and Weir 2006; Weir and Lunam 2004). These blood vessels presumably provide oxygen and nutrients by diffusion to the overlying superficial dermis and epidermis. The presence of this loosely arranged region in the superficial dermis probably exacerbates the above-mentioned loose grain defects in the tanned skin, and the blood vessels would render the skin susceptible to bruising. The presence of these superficial vessels provides anatomical evidence for a superficial blood supply, which may aid heat exchange with the external environment. Interestingly, these vessels are more dense in emu chicks at approximately 1 week of age compared to adult emus (Weir 2004). It is not known how this higher density of superficial vessels affects the ability of emu chicks to control heat and water loss across the skin. Despite the lack of knowledge on this issue, structural differences in the chick epidermis, differences in the vascular density, and the smaller size of chicks suggest that optimal conditions for the control of heat transfer and water loss and thus the welfare of young birds may be different to that of adults.

In both the ostrich and emu, the *stratum compactum* is a dense layer of connective tissue that consists predominantly of collagen (Lunam and Weir 2006; Weir and Lunam 2004). The most striking feature of this layer is that the collagen bundles are oriented parallel to the skin surface with few fibres running perpendicular to the epidermis (Fig. 7.3). In the ostrich, this has been observed using both light microscopy (Lange 1929; reviewed by Bezuidenhout 1999; Lunam and Weir 2006) and scanning electron microscopy (Fig. 7.3; Lunam and Weir 2006). This arrangement has also been described using both light microscopy (Frapple et al. 1997; Weir and Lunam 2004) and scanning electron microscopy (Weir 2004) in the emu. As in the *stratum superficiale*, the lack of collagen bundles running perpendicular to the skin surface may predispose to the lamination and loose grain defects seen in tanned skins. The paucity of elastic fibres suggests that both the strength and flexibility of the skin is due to the three-dimensional cross-weave arrangement of the collagen bundles.

By far, the thickest layer of skin in the emu is the *stratum laxum* (2.5 ± 1.1 mm in 18-month-old emus; Weir and Lunam 2004). This layer has not been described in the ostrich because the skin is separated between the stratum compactum and *stratum laxum* during flaying for the production of tanned skins. In feathered regions of the emu, the *stratum laxum* is filled with adipose tissue with some supporting collagen bundles, blood vessels, and nerves (Fig. 7.2; Weir and Lunam 2004). In the emu, the adipose tissue is removed for the production of emu oil. Feather follicles extend deep into the dermis and the base of the follicle sits deep within the *stratum laxum* (Fig. 7.2; Weir and Lunam 2004). Pennamotor muscles (*Mm. pennarum*) lie predominantly within the *stratum laxum*. These smooth muscles attach to the connective tissue sheath of the feather follicle either directly or via elastic fibres (Weir and Lunam 2004) and act to erect the feathers. Feather erection occurs in ostriches at high ambient temperatures (Crawford and Schmidt-Nielsen 1967; Luow et al. 1969; Phillips and Sanborn 1994), whereas the feathers flatten at low body temperature (Luow et al. 1969). Recently, Maloney (2008) suggested that feather erection may enhance convective and evaporative heat loss in the ostrich and emu. Therefore, these muscles may play an important role in thermoregulation.

7.3 Innervation of the Integument

The skin is also an important sensory organ and represents the first point of contact between the bird and its external environment. Knowledge of the sensory innervation of ratite skin (excluding the digits and beak) is largely derived from studies of the emu, but with limited information available from the ostrich. The skin adjacent to the wing of week-old emus is supplied by both sensory and sympathetic nerves (Weir 2004; Weir and Lunam 2006). Nerves are found in all dermal layers except the *lamina elastica* (Weir and Lunam 2006). Studies by Weir and Lunam (2006) have identified chemically distinct subpopulations of axons that target specific structures within the skin. Nerves have been observed within the superficial dermis, associated with the connective tissue sheath of feather follicles, within pennamotor muscles, and closely associated with the cutaneous vasculature (Weir and Lunam 2006). Free nerve endings and nerves with specialised end receptors are discussed in detail below. Included in the discussion are the implications of these nerves for the welfare of ratites.

7.3.1 Free Nerve Endings

The superficial dermis of the emu possesses free nerve endings immunoreactive for antibodies against the neuropeptides substance P (SP) and calcitonin gene-related peptide (CGRP) (Weir and Lunam 2006). Neural tracing studies have confirmed that a subpopulation of sensory neurons (within dorsal root ganglia) that project to the skin of the emu is immunoreactive for SP and CGRP (Weir 2004). While no physiological data are available in ratites, studies in mammals suggest that nerves containing SP and CGRP are involved in nociception (reviewed by Lawson 2002; Lu et al. 2003). These studies taken together support the existence of free nerve endings within the superficial layers of the emu skin that may be responsible for carrying impulses perceived as pain. The presence of these nerves within the superficial layers of the skin has implications for the welfare of emus because any abrasions or cuts could result in pain via stimulation of these nerves. Thus, care should be taken in all aspects of husbandry to minimise skin damage not only to produce optimal quality tanned skins, but more importantly, to reduce pain associated with skin injury.

7.3.2 *Herbst Corpuscles*

The Herbst corpuscle is a cutaneous sensory receptor that has been identified in volant birds as well as ostriches and emus. Its inner core is comprised of an encapsulated sensory nerve that is surrounded by several concentric layers of connective tissue. This tactile receptor detects vibration and is especially numerous in the beak of the domestic chicken (Lunam 2005). In feathered regions, Herbst corpuscles are invariably found adjacent to the follicles of the contour feathers (Dorward 1970; Hörster 1990; Lucas and Stettenheim 1972). Consistent with their distribution in the feathered integument of volant birds, Herbst corpuscles have been observed adjacent to the follicle sheath in the ostrich (Fig. 7.4; Lunam and Weir 2006), and in the emu they are found closely associated with the attachment of the pennamotor muscles to the feather follicle (Fig. 7.2; Weir and Lunam 2004). Similar receptors have also been reported in the plantar and digital pads of the

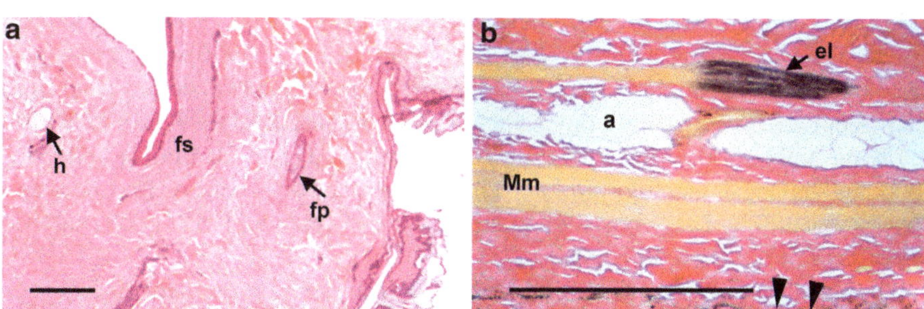

Fig. 7.4 Light micrographs of ostrich skin. (**a**) *h* Herbst corpuscle; *fs* follicle sheath; and *fp* sheath of filoplume. Both the Herbst corpuscle and filoplume lie close to the feather follicle. The epidermis is visible on the right-hand side of the micrograph. Haematoxylin and eosin stain. (**b**) Deep layers of the *stratum compactum*. *a* adipose tissue; *Mm* two pennamotor muscles running parallel in the *compactum*; and *e* elastic fibres of a pennamotor muscle. Elastic fibres (*arrows*) are visible among the dense collagen fibres. Verhoeff and van Gieson stain. *Scale bars* = 400 μm. Reproduced with permission from Lunam and Weir (2006) Rural Industries Research and Development Corporation publication No. 06/054

ostrich (Palmieri et al. 2003) and in the dorsal dermis of all digits of adult emus (Lunam and Glatz 2000). Although no physiological studies have been conducted in ratites, the location of Herbst corpuscles adjacent to the feather follicle suggests that ostriches and emus can detect movement of the feathers via these receptors.

7.4 Filoplumes and Bristle Hairs

A discussion of the structure of ratite skin would be incomplete without the mention of filoplumes and bristle hairs. These miniature feathers are unique amongst all feathers in that they lack attachment of smooth muscles to their follicle sheath. A detailed account of the structure of these tiny feather-like projections can be found in a review by Lucas and Stettenheim (1972). Although minor variations in the macroscopic and microscopic appearance of filoplumes and bristle hairs have been reported amongst numerous avian species, fully mature filoplumes and avian bristles invariably consist of a slender rachis, which differ in the shape of the shaft, and the presence or absence of a tuft of apical barbs. In contrast to Lucas and Stettenheim (1972), who reported the absence of filoplumes in the feathered regions of ratites, both filoplumes and bristle hairs have been identified in the ostrich skin (Fig. 7.5; Lunam and Weir 2006; Glatz 2010). Filoplumes of the adult ostrich have a structure similar to the late immature filoplumes of the domestic chicken (Lucas and Stettenheim 1972). They consist of a single tapered rachis surrounded by 2–12 barbs extending 8–12 mm in length (Fig. 7.5). Barbules are present along the extent of the shaft of the barbs. Distal barbules are slightly longer than proximal barbules giving a tuft-like appearance at the tip of the barbs.

Scanning electron microscopy revealed structures resembling bristle hairs in the crown region of ostrich skin (Lunam and Weir 2006). These hair-like feathers have a cluster of two to four short barbs at the base of a single rachis (3–5 mm in length) with an absence of barbs at the tip (Fig. 7.5; Lunam and Weir 2006). Further support for the existence of filoplumes and bristle hairs in the ostrich skin comes from light microscopic studies revealing tiny follicle sheaths (approximately 50 μm in diameter) lying within 500 μm of a contour feather follicle (Fig. 7.4; Lunam and Weir 2006) and having a histology similar to filoplume sheaths in other Aves (Lucas and Stettenheim 1972). They consist of an inner lining of epidermal cells encircled by a dense band of collagen fibres.

The function of avian bristles is speculative and is based on their morphology and location. Narial bristles are considered to filter air and prevent foreign materials from entering the nostrils, while those on the eyelids protect the eyes and act as proprioceptors. Bristles on other regions of the head likely serve as tactile receptors. According to Lucas and Stettenheim (1972), avian bristles are almost exclusively found in the head region and are not associated with the contour feathers.

Assumptions regarding the function of filoplumes, similar to the case for avian bristles, are based on their morphology and location within the skin. To our knowledge, no physiological studies have been conducted to definitively determine

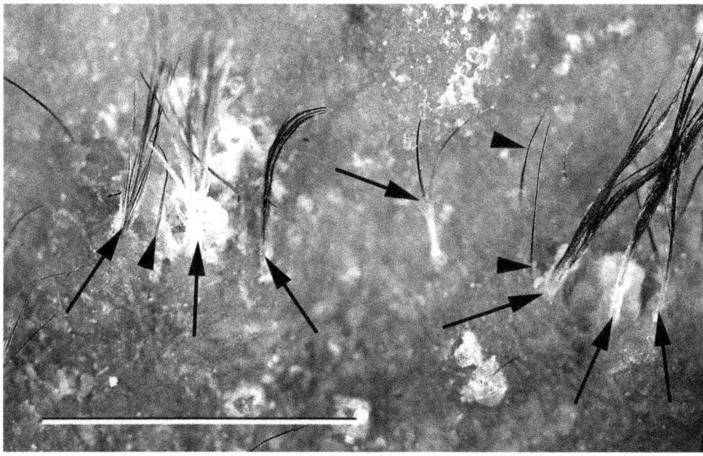

Fig. 7.5 Raw salted ostrich skin. Filoplumes (*arrows*) are arranged in a semicircle on the caudal side of two feather follicles. Interspersed between the filoplumes are bristle hairs (*arrowheads*). A filoplume (*arrow* indicates shaft) and a bristle hair (*arrowhead* at shaft) can be seen in the interfollicular space. Filoplumes and bristle hairs were found at the base of the majority of feather follicles. Crystals of hide salt are visible as *white* material on the skin surface. *Scale bar* is 1 cm. Reproduced with permission from Lunam and Weir (2006) Rural Industries Research and Development Corporation publication No. 06/054

their role in feathered skin. As they are found in close proximity to the feathers and Herbst corpuscles, Lucas and Stettenheim (1972) have suggested they may serve as tactile receptors and transmit slight changes in movement of the contour feathers (detected via the follicle sheath) to nearby Herbst corpuscles. In the ostrich, as in other Aves, filoplumes are clustered at the base of the contour feathers. Interestingly, in our study in the ostrich, filoplumes and bristle hairs had the same distribution pattern. Furthermore, in the ostrich they form a semicircular pattern exclusively on the caudal side at the base of the contour feathers (Figs. 7.5 and 7.6; Lunam and Weir 2006; Black and Glatz 2008; Glatz 2010). The contour feathers are angled in a cranial to caudal direction directly above the filoplumes and bristle hairs. Consequently, during movement, the feathers are likely to brush against and bend the distal rachis of the filoplumes and bristle hairs. This particular arrangement tempts us to speculate that filoplumes and possibly bristle hairs in the feathered skin of the ostrich may not provide indirect sensory information via the Herbst corpuscles as previously suggested, but are direct sensory receptors that convey spatial information to the central nervous system on the position of the contour feathers.

Lucas and Stettenheim (1972) and Borodulina (1966) reported innervation of the follicular wall of the contour feathers. This is in contrast to the results of Ostmann et al. (1963) who report an absence of nerve fibres within the follicular sheath of the domestic chicken. Our studies in the emu have demonstrated nerve fibres surrounding the follicular sheath of contour feathers that are immunoreactive for antibodies against substance P, calbindin D-28k, or neuropeptide Y (Weir and Lunam 2006).

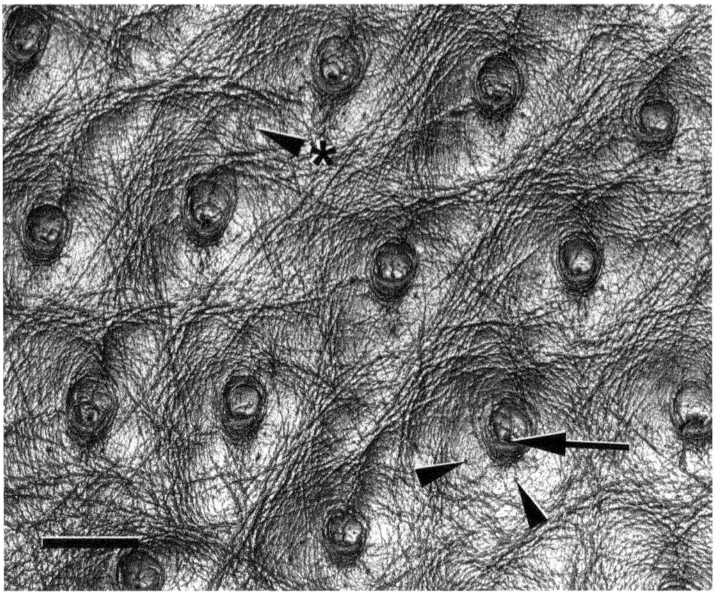

Fig. 7.6 Crown area of chrome-tanned ostrich skin. The holes (*arrow*) from the shafts of the contour feathers are located in the caudal region of the raised follicles. The follicles are orientated to accommodate the contour feathers angled in a cranial to caudal direction. Immediately caudal to the base of each follicle lie several discrete pinholes (*arrowheads*). The skin is noticeably plumped in the region of the pinholes at the caudal side of each follicle. An interfollicular pinhole is visible (*arrowhead* with *asterisk*). *Scale bar* is 1 cm. Reproduced with permission from Lunam and Weir (2006) Rural Industries Research and Development Corporation publication No. 06/054

The dermal pulp of most feather follicles in the domestic chicken is reported to contain nerve fibres immunoreactive for substance P and some contain calbindin D-28k-positive nerve fibres (Duc et al. 1993). We are attempting to elucidate the relationship between Herbst corpuscles, filoplumes, and contour feathers by identifying the neural connections and further classifying the sensory nerves innervating these structures.

In the ostrich, filoplumes and bristle hairs are also found interspersed between the contour feather follicles. The distribution and density of these interfollicular structures are highly variable within ostriches of a given flock as well as between different strains (Lunam and Weir 2006; Glatz 2010). Filoplumes and bristle hairs have identical distributions for a given skin (Lunam and Weir 2006), supporting anecdotal suggestions within the ostrich industry of a genetic heritability of these traits.

Removal of filoplumes results in tiny pinholes in tanned ostrich skins. Cooper (2001) has cautioned against confusing these "true" pinholes with random defects in the skin. In our study, the pattern of pinholes in tanned skins paralleled the distribution of filoplumes and bristle hairs (Lunam and Weir 2006). This suggests that the pinholes result from removal of the filoplumes and bristle hairs and are not incidental defects in the skins caused by bacterial contamination or mechanical trauma. Interfollicular pinholes in the crown region significantly downgrades tanned ostrich skins, thereby reducing economic return. This reduction in skin quality has sparked research into the heritability of filoplumes in the ostrich skin in an attempt to eradicate the interfollicular filoplumes.

It is not known whether these interfollicular filoplumes and bristle hairs have a proprioceptive function similar to those associated with the contour feathers. However, their highly variable distribution among individual ostriches suggest that selective breeding to remove these interfollicular structures is unlikely to have any welfare impact by reducing tactile sensation in the skin. The concept of selective breeding to remove interfollicular filoplumes and bristle hairs poses several questions. How genetically linked are the filoplumes at the base of the contour feathers to interfollicular filoplumes? What is the genetic link between bristle hairs in the feathered skin to those serving as proprioceptors around the eyes and ears of the ostrich? As the structure of filoplumes and bristle hairs is highly conserved in Aves, how essential are they in proprioception and how resistant are these traits to genetic manipulation?

7.5 Concluding Remarks

The anatomy of the integument of the ostrich and the emu has been described at the light and electron microscopic levels. Attempts at understanding the physiological function of these structures in ratites have largely arisen from comparison to work conducted in volant birds. To achieve a greater understanding of the role of the skin on the welfare of ratites, we have highlighted a number of areas for further study. The presence of several specialised receptors and potential nociceptors innervating the skin suggests that it is a highly specialised organ capable of detection of pressure, touch, and feather movement as well as transmission of painful stimuli. Therefore, during agistment of commercially farmed birds, care needs to be taken to minimise any damage to the skin and thereby avoid compromising the welfare of the birds. How the age-related changes in the skin and feathers affect the ability of the bird to interact with its environment is poorly understood. Research is required to ensure optimal husbandry practices during growth from young chicks until adulthood. In addition, further work is required to determine the role of the epidermis in thermoregulation. This phenomenon is of particular interest in understanding how ratites have adapted to survival in hot arid climates. Finally, in addition to the promotion of commercial interests in improving the quality of tanned skins, future studies are required to examine the potential mechanoreceptive role of filoplumes and bristle hairs in the ostrich.

Acknowledgements The authors' research was supported by grants from the Australian Research Council and the Rural Industries Research and Development Corporation.

References

Aszterbaum M, Menon GK, Feingold KR, Williams ML (1992) Ontogeny of the epidermal barrier to water loss in the rat: correlation of function with stratum corneum structure and lipid content. Pediatr Res 31:308–317

Bezuidenhout AJ (1999) Anatomy. In: Deeming DC (ed) The ostrich biology, production and health. CABI Publishing, London, pp 13–49

Black D, Glatz PC (2008) Filoplumes and pinholes in ostrich hides (Abstract). Worlds Poult Sci J 64 (Suppl 2, XXIII):661

Borodulina TL (1966) The innervation of the filoplumes. In: Kleinberg SE (ed) Mechanisms of flight and orientation of birds. Akad. Nauk, SSSR, Inst Morfologii Zhivotn, Moscow, pp 113–145

Cooper RG (2001) Ostrich (*Struthio camelus var. domesticus*) skin and leather: a review focused on southern Africa. Worlds Poult Sci J 57:157–178

Crawford EC Jr, Schmidt-Nielsen K (1967) Temperature regulation and evaporative cooling in the ostrich. Am J Physiol 212:347–353

Dorward PK (1970) Response patterns of cutaneous mechanoreceptors in the domestic duck. Comp Biochem Physiol 35:729–735

Duc C, Barakat-Walter I, Droz B (1993) Calbindin D-28k- and substance P-immunoreactive primary sensory neurons: peripheral projections in chick hindlimbs. J Comp Neurol 334:151–158

Elias PM, Friend DS (1975) The permeability barrier in mammalian epidermis. J Cell Biol 65: 180–191

Frapple P, O'Malley P, Snowden J, Hagan R (1997) Emu processing and product development. Rural Industries Research and Development Corporation Research Publication No. 97/66

Glatz PC (2010) Husbandry and genetic strategies to improve hide quality of ostriches. Rural Industries Research and Development Corporation Publication No. 09/174

Glatz PC, Miao ZH (2008) Husbandry of ratites and potential welfare issues: a review. Aust J Exp Agric 48:1257–1265

Hörster W (1990) Histological and electrophysiological investigations on the vibration-sensitive receptors (Herbst corpuscles) in the wing of the pigeon (*Columba livia*). J Comp Physiol A 166: 663–673

Lange B (1929) Uber enige besondere formen des faserverlaufes im bindegewebe der vogelhaut. Anat Anz 67:452–459

Lawson SN (2002) Phenotype and function of somatic primary afferent nociceptive neurones with C-, Aδ- or Aα/β-fibres. Exp Physiol 87:239–244

Lu Y, Park T, Rice FL, Laurito CE (2003) Absence of substance P and CGRP in the dorsal root ganglia of naked mole rats correlates with an absence of hyperalgesia to heat. In: Proceedings of the 10th World Congress on Pain, Progress in Pain Research and Management, vol 24. pp 227–234

Lucas AM, Stettenheim PR (1972) Avian anatomy – integument (Parts I & II). United States Department of Agriculture, Washington, DC

Lunam CA (2005) The anatomy and innervation of the chicken beak: effects of trimming and retrimming. In: Glatz PC (ed) Poultry welfare issues: beak trimming. Nottingham University Press, Nottingham, UK, pp 51–68

Lunam CA, Glatz PC (2000) Declawing of farmed emus; harmful or helpful? Rural Industries Research and Development Corporation Publication No. 99/177

Lunam CA, Weir KA (2006) Storage of ostrich skin: effects of preservation methods on skin structure, physical properties and microbial flora. Rural Industries Research and Development Corporation Publication No. 06/054

Luow GN, Belonje PC, Coetzee HJ (1969) Renal function, respiration, heart rate and thermoregulation in the ostrich (*Struthio camelus*). Sci Pap Namib Desert Res Station 42:43–54

Maloney SK (2008) Thermoregulation in ratites: a review. Aust J Exp Agric 48:1293–1301

Maloney SK, Dawson TJ (1994) Thermoregulation in a large bird, the emu (*Dromaius novaehollandiae*). J Comp Physiol B 164:464–472
Maloney SK, Dawson TJ (1998) Changes in the pattern of heat loss at high ambient temperature caused by water deprivation in a large flightless bird, the emu. Physiol Zool 71:712–719
Menon GK, Menon J (2000) Avian epidermal lipids: functional consideration and relationship to feathering. Am Zool 40:540–552
Menon GK, Brown BE, Elias PM (1986) Avian epidermal differentiation: role of lipids in permeability barrier formation. Tissue Cell 18(1):71–82
Menon GK, Baptista LF, Elias PM, Bouvier M (1988) Fine structural basis of the cutaneous water barrier in nestling Zebra Finches *Poephila guttata*. Ibis 130:503–511
Menon GK, Maderson PFA, Drewes RC, Baptista LF, Price LF, Elias PM (1996) Ultrastructural organisation of avian stratum corneum lipids as the basis for facultative cutaneous waterproofing. J Morphol 27:1–13
Ostmann OW, Ringer RK, Tetzlaff M (1963) The anatomy of the feather follicle and its immediate surroundings. Poult Sci 42:958–969
Palmieri G, Sanna M, Minelli SB, Botti M, Gazza F, Di Summa A, Santamaria N, Passantino L, Maxia M, Acone F (2003) On the sensitive innervation of the ostrich's foot pads. Ital J Anat Embryol 108(1):25–37
Phillips PK, Sanborn AF (1994) An infrared, thermographic study of surface temperature in three ratites: ostrich, emu and double-wattled cassowary. J Therm Biol 19:423–430
Ross MH, Pawlina W (2006) Histology: a text and atlas with correlated cell and molecular biology. Lippincott Williams and Wilkins, Philadelphia
Schmidt-Nielsen K, Kanwisher J, Lasiewski RC, Cohn JE, Bretz WL (1969) Temperature regulation and respiration in the ostrich. Condor 71(4):341–352
Weir KA (2004) The microanatomy of emu skin and implications for the Australian emu industry. PhD thesis, Flinders University, Adelaide, p 147
Weir KA, Lunam CA (2004) A histological study of emu (*Dromaius novaehollandiae*) skin. J Zool (Lond) 264:259–266
Weir KA, Lunam CA (2006) Immunohistochemical study of cutaneous nerves in the emu. Cell Tissue Res 326:697–705
Withers PC (1983) Energy, water, and solute balance of the ostrich Struthio camelus. Physiol Zool 56(4):568–579

Chapter 8
Ratite Movement

R.G. Cooper

Abstract The ratites evolved specialised for terrestrial locomotion and as a consequence are rapid moving birds, and this has helped the surviving species to partially avoid predation and danger. Where possible, welfare issues were highlighted and suggestions were made for future investigations. Ratites are believed to be monophyletic, with the flighted tinamous as their sister group, suggesting a single loss of flight in the common ancestry of ratites. Evolutionary considerations of movement also necessitate consideration of adaptations that facilitate foraging and feeding. The dynamics of ostrich locomotion and predicted model for the ecological existence of the ostrich could be ascertained by comparison with the extinct terror bird. Bow leg syndrome in ostrich chicks and tibiotarsal rotation in 14-month-old birds was investigated. More studies of locomotion and limb abnormalities in live emus, cassowaries, rheas, kiwis and tinamous are needed. Other important aspects of movement include the relationships between the dynamics of feeding and locomotion. Many more studies are needed to complete aspects of ratite locomotion in the wild during various activities (foraging, reproductive displays, running, defence and sleeping). More extensive research is needed on the various stages of embryonic limb development in ratites to determine whether a developmental, genetic or incubational influence is predisposing weakening of the osteoblastic and osteoclastic associations. Given the global threats to wild ratite habitat, more studies need to be performed to determine the precise nature of their niches (wild, ranch or farming) to reduce the real threat of complete extinction.

Keywords Ecology · Evolution · Feeding · Locomotion · Ratites · Running · Welfare

R.G. Cooper
Flat 5, Eurohouse, Dog Kennel Lane, Walsall WS1 2BU, England, UK
e-mail: rgcooperuk@yahoo.com

DOI 10.1007/978-3-642-19297-5_8,

8.1 Introduction

The ratites evolved specialised for terrestrial locomotion and as a consequence are rapid moving birds, and this has helped the surviving species to partially avoid predation and danger. Pre-historic ratites moved even faster, particularly the terror bird (Phorusrhacinae gen.) (Cooper and Tennett 2008a; Cooper 2008). Welfare considerations of the bird become extremely significant when a patho-morphological problem arises in the limbs resulting in severe, overwhelming effects on locomotion and dramatically attenuated mobility, inadequate feeding and susceptibility to predation. Although in captivity sophisticated locomotion problems can be thoroughly studied, more work is needed in the wild ratites particularly from an ecological and conservation perspective. More information about ratite locomotion needs to be made available in biological studies. Products associated with locomotion including the interesting back-scratchers for sale in Australian cultural stores made of the dried feet of emus and ostriches. Much work has been done to describe the anatomical and skeletal morphology of the limbs, feet and claws of the ratites and it is not the place of this chapter to repeat the like.

Hallam (1992) gave a summarised account of limb deformities in ostriches from observations made in Zimbabwe, and suggested that this has a strong bearing on chick mortality. Twisted or rotated leg (tibiotarsal rotation) occurs mostly at 2 weeks and 4–7 weeks (Chapter 9, Hallam 1992) with one leg usually rotated outwards with the foot twisted at least 90° to the anterior/posterior axis of the body (Hallam 1992). Bow leg syndrome occurs mostly at 6–8 weeks, commonly observed as bone curvature frequently in both legs (Hallam 1992). True rickets occurs in chicks aged less than 6 weeks and the bone is usually soft, sometimes with bowing and rotation (Hallam 1992). Horbańczuk (2002) described curl toes syndrome in which afflicted ostrich chicks avoid walking and spend much of their time sitting. A case of lameness in an ostrich aged 7 months was reported by Samorek-Salamonowicz et al. (2002). Exudates of a gelatinous consistency were observed in the ankle joint. Application of Soluvit multivitamins in the drinking water for 3 days resulted in a disappearance of lameness in a bird (Samorek-Salamonowicz et al. 2002). Patterns of limb deformity are clearly multi-factorial with the rapid increase in metatarsal length of 2 cm/week playing a role (Hallam 1992). Theoretical elaborations of probable causes are discussed as congenital, nutritional, traumatic, infection and lack of exercise (Hallam 1992). In ostrich chicks, further aspects of nutritional problems are discussed (Cooper 1999, 2000, 2002; Cooper and Horbańczuk 2004). Pathological indicators on dissection revealed bruising and detachment of the periosteum in the growth areas, varying degrees of mineralization and other facets (Hallam 1992). Congenital limb deformities are common. Fractured legs are extremely difficult to treat because of rapid leg growth and huge supportative forces (Hallam 1992). If nutrition of breeders and incubation parameters are correct, the treatment is usually ineffective and may include attempts at hobbling the legs (rope tied above the feet), sling support, support on a padded block, vitamin E/Selenium and vitamin C injection and supplementation of feed and water (2 mg/10 L manganour sulphate, 1 g/kg vitamin B/C and crushed, sterilised bone) (Hallam 1992). Overcrowding of chicks may pre-dispose leg aberrations and Madeiros (2007) tabulated the mandatory area per bird: for example, for ostrich, emu and rhea chicks aged less than 4 days, the maximum birds per group is 12/1 m^2 of shelter (0.25 m^2/bird).

The aim of this chapter is to perform a comprehensive literature search using articles from PubMed/ISI-indexed article searches, proceedings of conferences, reports, magazines, reprint requests and other sources, of important aspects of ratite locomotion. Where possible, welfare issues were highlighted and suggestions were made for future investigations.

8.2 Relevant Evolutionary Considerations

Ratites (ostrich, emu, rhea, cassowary, kiwi and tinamous) are large, flightless birds that have long fascinated biologists (Harshman et al. 2008). Their current distribution on isolated southern land masses is believed to reflect the break-up of the paleocontinent of Gondwana. Man and pests introduced by him have been responsible for the tragic loss of some impressive ratites including the giant elephant bird (500 kg b.wt.) in Madagascar and the giant moa (250 kg b.wt.) in New Zealand (Madeiros 2007). Many woody species in the thickets of southern Madagascar share, with New Zealand, anachronistic structural defences against large extinct bird browsers (Bond and Silander 2007). In addition, gastrolith clusters of some derived theropod dinosaurs (oviraptorosaurs and ornithomimosaurs) compare well with birds, suggesting that the gastric mill evolved in the avian stem lineage (Wings and Sander 2007).

Ratites are believed to be monophyletic, with the flighted tinamous as their sister group, suggesting a single loss of flight in the common ancestry of ratites. However, phylogenetic analyses of 20 unlinked nuclear genes revealed a genome-wide signal that unequivocally places tinamous within ratites, making them polyphyletic and suggesting multiple losses of flight (Harshman et al. 2008). The most plausible hypothesis requires at least three losses of flight and explains the many morphological and behavioural similarities among ratites by parallel or convergent evolution (Harshman et al. 2008). Phylogeny demands fundamental reconsideration of proposals that relate ratite evolution to continental drift (Harshman et al. 2008).

Evolutionary considerations of movement also necessitate consideration of adaptations that facilitate foraging and feeding. Several different types of cranial kinesis are present within modern birds, enabling them to move (part of) the upper bill relative to the braincase (Gussekloo et al. 2001). This movement of the upper bill results from movement of the quadrate and the pterygoid–palatine complex (PPC). The taxon Palaeognathae is characterised by a very distinct PPC and a special type of cranial kinesis (central kinesis) that is very different from that in the Neognathae. A Roentgen stereophotogrammetry method was used to measure bone movements. A new method was developed to quantify motion or deformation of bony structures as quantification is often difficult due to overlaying tissue, and the currently used roentgen stereophotogrammetry method requires significant investment (Gussekloo et al. 2000). A single stationary roentgen source was used, as opposed to the usual two, which, in combination with a fixed radiogram cassette holder, forms a camera with constant interior orientation. By rotating the experimental object, it is possible to achieve a sufficient angle between the various viewing directions, enabling photogrammetric calculations. Co-ordinates of spherical markers in the head of *Rhea americana* were calculated with an accuracy of 0.12 mm. When these co-ordinates were used in a deformation analysis, relocations of ca. 0.5 mm could be accurately determined (Gussekloo et al. 2000). External forces applied during food acquisition may influence the morphology of the PPC. Differences in PPC morphology therefore appear to be the result of different functional demands acting on the system simultaneously but with different strengths, depending on the species (Gussekloo et al. 2001). This study could be usefully extended to consider neck, torso and leg segmental changes between the extinct and within the surviving ratites. In the extinct carnivorous terror bird (Phorusrhacinae gen.), strength of limbs not only allowed to capture large prey, but also assisted in warding off predators from kills and breaking bones to reach the nutritious bone marrow (Cooper and Tennett 2008a; Cooper 2008). In addition, apart from a defence tool, did the claws evolve to allow the ratites to dig to a certain depth in soil while seeking food or constructing a nest and what is the histology of the claw enabling its strength? More podiatric investigations of foot morphology, function, movement and impediment are needed in ratites.

8.3 Surviving Ratites: Locomotion and Welfare Issues

*8.3.1 Ostrich (*Struthio camelus*)*

Cooper and Tennett (2008a) and Cooper et al. (2008a, b, c) proposed mathematical formulae to calculate the dynamics of ostrich locomotion. The authors also proposed a model to predict the ecological existence of the ostrich by comparison with the extinct terror bird. From an environmental and welfare perspective, formulae for locomotion are essential for comparing locomotion characteristics of the living ratites (ostrich, emu, rhea, cassowary, kiwi and tinamou) and extinct flightless birds (moa, dodo and elephant bird).

Studies of bow leg syndrome in ostrich chicks aged 2, 4, 8 and 12 weeks incorporated detailed measurements of femur plus tibiotarsus; tarsometatarus; phalanx I, digit III; phalanx II, digit III plus phalanx III, digit III; and phalanx IV, digit III (Cooper 2007a; Cooper et al. 2008a, b, c) (Fig. 8.1). To determine stride and speed, a sand run was constructed (6 $\times$ 1.7 m) and subdivided into 2 m sections and the time taken to traverse it was recorded (Cooper et al. 2008a, b, c). Measurements (cm) were made of the left and right footprints, the number of footprints and

Fig. 8.1 Experimental run with ostrich chicks in the Egyptian desert

average stride length in 0–2, 3–4 and 5–6 m (Cooper et al. 2008a, b, c). The number of steps was greater in bow leg chicks aged 4 and 8 weeks than in healthy birds. However, stride length was smaller in all age groups with bow leg. All speeds in bow leg chicks were lower than those in healthy birds, except for that recorded at 2 m in chicks aged 2 weeks, which did not differ markedly (Cooper et al. 2008a, b, c). In affected birds, feathers were sparse and icterus was present. The tarsometatarsus was twisted, with severely inflamed joints, eroded distal ends, thickening of the cartilage and the presence of fibrous material surrounding the ligaments (Cooper et al. 2008a, b, c). Muscles in the hind limb were emaciated. The authors suggested that the syndrome compromises the ability of chicks to keep up with adults in flocks, and may with lessen their ability to escape predation (Cooper et al. 2008a, b, c).

Tibiotarsal rotation was investigated in 20 ostriches (9 cocks and 11 hens) (14-month-old) in a run (50 $\times$ 2.5 m), which was divided into sections marked 5, 10, 15 and 20 m (Cooper 2007b) (Fig. 8.2). The degree of tibiotarsal rotation in the right foot was 156 (mean) $\pm$ 2.69°. Comparisons between left and right foot length in healthy birds showed no significant differences. Foot length was significantly lower in tibiotarsal rotation ($P = 0.03$). The right foot in tibiotarsal rotation was significantly shorter than the left foot. The number of strides per each 5 m division were significantly ($P < 0.05$) greater in tibiotarsal rotation by comparison with healthy birds (Cooper 2007b). At 20 m, healthy cocks had more strides than hens. The stride length in hens was significantly ($P < 0.05$) greater than cocks at 5, 10 and 15 m, but lower throughout in tibiotarsal rotation ($P = 0.001$). The speed of hens was significantly ($P < 0.05$) greater than cocks (Cooper 2007b). Tibiotarsal rotation resulted in significantly ($P < 0.05$) reduced speeds. Hens may be able to escape

Fig. 8.2 Experimental run with ostrich breeders in Poland

danger faster than cocks. The occurrence of tibiotarsal rotation necessitates consideration of genetics, management, sex, nutrition and growth rates (Cooper 2007b).

This study was followed by the development of a novel method of digital analysis of a radiological image of the tibiotarsal bone of ostriches (Charuta et al. 2008). Using a resolution of 0.096 mm/pixel, selective images of rectangular-shaped areas (160 × 320 points) of spongy bone was taken ca. 80 mm posterior to the articular surface near the proximal metaphysis. Although there was no significant difference in trabeculation between hens and cocks, the method is very useful for determining bone densities in leg deformities and eliminates the need to kill the chick.

Ostriches increase their speed by increasing the frequency of strides and their centre of mass is close to the hip, and their hind limbs have an extended jointed chain system with a short erect femur, maximising a gravity-powered system (Abourachid and Renous 2000). In this study, step length was associated with the protraction and the retraction angles.

Schaller et al. (2005) used a morphometric analysis of the pelvic and hind limb skeleton of an ostrich in conjunction with surveys of 83 ratite skeletons, and dissection of the locomotor apparatus of *Struthio camelus*, *Pterocnemia pennata* and *Dromaius novaehollandiae* and determined muscle distribution and mass. Functional assessment of the metatarsus demonstrated that the ostrich exhibited optimal leg segment lengths. This would presumably aid in the power dynamics of their locomotion.

Muscle moment arms were measured for major muscles of the pelvic limb of the ostrich (*S. camelus*) to assess specific functional behaviour and to apply this to locomotor performance (Smith et al. 2007). The tendon travel technique was used to measure moment arms of 21 muscles at the hip, knee, ankle and metatarsophalangeal joints throughout the ranges of motion observed during level running. Moment arm lengths tended to be the longest for the large proximal muscles, while the largest relative changes were found for the moment arms of the distal muscles. The greatest relative changes were observed in the flexors of the metatarsophalangeal joint, for which a threefold increase in moment arm was observed from flexion to full extension. Changes in muscle moment arm through the range of motion studied appear to optimise muscle function during stance phase, increasing the effective mechanical advantage of these muscles (Smith et al. 2007).

Although locomotor kinematics in walking and running birds have been examined in studies exploring many biological aspects of bipedalism, these studies have been largely limited to two-dimensional analyses (Rubenson et al. 2007). A five-segment, 17 degree-of-freedom kinematic model of the ostrich hind limb was developed from anatomical specimens to determine the three-dimensional joint axis alignment and joint kinematics in the ostrich during running (ca. 3.3 m/s). The majority of the segment motion during running in the ostrich occurred in flexion/extension. The alignment of the average flexion/extension helical axes of the knee and ankle are rotated externally to the direction of travel (37° and 21°, respectively), so that, pure flexion and extension at the knee will act to adduct the tibiotarsus relative to the plane of movement, and pure flexion and extension at the ankle will act to abduct and adduct the tarsometatarsus relative to the plane of movement (Rubenson et al. 2007). This feature of limb anatomy appeared to provide the major lateral (non-sagittal) displacement of the lower limb necessary for steering the swinging limb clear of the stance limb and replaces what would otherwise require greater adduction/abduction and/or internal/external rotation, allowing for less complex joints, musculoskeletal geometry and neuromuscular control. Hip abduction and knee internal/external and varus/valgus motion may further facilitate limb clearance during the swing phase, and substantial non-flexion/extension movement at the knee is also observed during stance (Rubenson et al. 2007).

Eight ostriches were induced to run along a trackway and execute turns (Jindrich et al. 2007). Ostriches executed manoeuvres using a simple control strategy that required minimal changes to leg kinematics or net torque production at individual joints. Although ostriches did use acceleration or braking forces to control body rotation, their morphology allowed for both crossovers and sidesteps to be accomplished with minimal net acceleratory/braking force production. Body roll and abduction/adduction of the leg shifted the foot position away from the turn direction, reducing the acceleratory/braking forces required to prevent under- or over-rotation and aligning the leg with the ground reaction force (Jindrich et al. 2007).

The functional anatomy of the pelvic limb of the ostrich (*S. camelus*) was investigated to assess musculoskeletal specialisation related to locomotor performance (Smith et al. 2006). The pelvic limbs of ten ostriches were dissected and detailed measurements of all muscle tendon units of the pelvic limb were made, including muscle mass, muscle length, fascicle length, pennation angle, tendon mass and tendon length. Larger muscles tended to be located more proximally and had longer fascicle lengths and lower pennation angles (Smith et al. 2006). Generally, high-power muscles were located more proximally in the limb, while some small distal muscles (*tibialis cranialis* and *flexor perforatus digiti III*), with short fibres, were found to have very high force generation capacities (Smith et al. 2006). The greatest proportion of pelvic muscle volume was for the hip extensors, while the highest capacity for force generation was observed in the extensors of the ankles, many of which were also in series with long tendons, and thus, were functionally suited to elastic energy storage (Smith et al. 2006).

Biomechanical and metabolic measurements of ostriches moving on a treadmill over a range of speeds from 0.8 to 6.7 m/s showed that the selection of walking or grounded running at intermediate speeds favoured a reduction in the metabolic cost of locomotion (Rubenson et al. 2004). This gait transition was characterised by a shift in locomotor kinetics from an inverted pendulum gait to a bouncing gait that lacked an aerial phase (Rubenson et al. 2004). By contrast, when the ostrich adopted an aerial-running gait at faster speeds, there were no abrupt transitions in mechanical parameters or in the metabolic cost of locomotion. These data suggested a continuum between grounded and aerial running, indicating that they belonged to the same locomotor paradigm (Rubenson et al. 2004).

The gross morphology and the flexibility along the neck of the ostrich (*S. camelus*) were examined using fresh tissue as well as neck skeletons (Dzemski 2007). The neck of the ostrich can be divided into four sections with different functions: the atlas–axis complex, which is responsible for torsion; the adjacent cranial section of the neck is flexible in dorsoventral and lateral directions, but this part of the neck is usually kept straight at rest and during feeding; dorsoventral flexibility is highest in the middle section of the neck; and the base of the neck is primarily used for lateral excursions (Dzemski 2007).

Haematoxylin–eosin or different gold chloride impregnations were used to detect the sensitive and autonomic innervation of foot pads (Palmieri et al. 2003). The autonomic innervation was represented by isolated or grouped ganglion cells located along the course of nerve bundles. The capsulated nerve endings, morphologically classified as Pacini, Pacini-like and Herbst corpuscles, were not uniformly distributed throughout the considered districts and their number was always higher in the plantar pad compared with the digital pads (Palmieri et al. 2003). Further studies should be carried out to determine the musculo-synaptic associations of muscle contractility and ionic movement.

On the basis of behavioural observations of breeding ostriches, a study evaluated ostrich housing under South German climatic conditions (Rhine level) (Wöhr et al. 2005). The locomotion activity of the animals was strongly associated with the reproductive periods and the territorial behaviour and therefore highest in spring. On cold days, the animals performed their reproduction behaviour mainly in the stable. For the comfort behaviour, distinct weather dependence was seen particularly for sand bathing. Warm temperatures and dry sand were the preconditions for it (Wöhr et al. 2005). Indeed, more studies are required on farms and compared with wild ostriches to determine changes in locomotion dynamics. Populations of wild ostriches exist in certain areas of Africa including southern Africa (Fig. 8.3a, b).

Fig. 8.3 (**a**) Wild ostrich near coast of South Africa (courtesy of D. Hargrove). (**b**) Mature ostrich growers on an Egyptian farm

*8.3.2 Emu (*Dromaius novaehollandiae*)*

Emus live in most habitats across Australia, although they are most common in areas of sclerophyll forest and savannah woodland, and least common in populated and very arid areas. Emus are largely solitary, and while they can form enormous flocks, this is an atypical social behaviour that arises from the common need to move towards food sources. Emus forage in a diurnal pattern. They eat a variety of native and introduced plant species; the type of plants eaten depends on seasonal availability. They also eat insects, including grasshoppers and crickets and other beetles and larvae. Emus serve as an important agent for the dispersal of large viable seeds, which contributes to floral biodiversity. Emus form breeding pairs during the summer months of December and January, and may remain together for about 5 months. Mating occurs in the cooler months of May and June.

Emus provide an excellent opportunity for studying the sustained high-speed running by a bird (Fig. 8.4a, b) (Patak and Baldwin 1998). As emus evolved, selection was directed towards their limbs with the attenuation of their wing skeleton to only a single functional digit (Maxwell and Larsson 2007). Their pelvic limb musculature is described in detail and morphological features and characteristics of a cursorial lifestyle are identified. Several anatomical features of the pelvic limb reflect the emus' ability for sustained running at high speeds: emus have a reduced number of toes and associated muscles; have a unique *Muscularis gastrocnemius* with four muscle bellies, not the usual three; and contribution to total body mass of the pelvic limb muscles of emus is similar to that of the flight muscles of flying birds (Patak and Baldwin 1998). In general, the pelvic limb musculature of emus resembles that of other ratites with the notable exception of *M. gastrocnemius*. The presence and arrangement of four muscle bellies may increase the effectiveness of *M. gastrocnemius* and other muscles during cursorial locomotion by moving the limb in a craniocaudal rather than a lateromedial plane (Patak and Baldwin 1998).

Fig. 8.4 (**a**) The author coaxing an emu in Brisbane, Australia to sit. (**b**) Emu in Koala Centre, Brisbane, Australia

Kinematic and kinetic data captured for two laboratory-habituated emus showed that the ground reaction forces had a bimodal distribution over the course of the stance phase (Goetz et al. 2008). Two bird-averaged maximum hip contact force was ca. 5.5 × b.wt., directed nominally axially along the femur (Goetz et al. 2008).

Orthopaedic management of femoral head osteonecrosis remains problematic, partly because of inability to systematically compare treatments in an animal model whose natural history parallels the human in terms of progression to femoral head collapse (Troy et al. 2007). Recently, it was determined that collapse could be consistently achieved for cryogenically induced osteonecrosis in the emu. The average number of steps taken per day was 9,563, which extrapolates to 1.8 million hip loadings per year (Troy et al. 2007). On average, the emus spent 4:05 h/day idly standing, 2:12 h squatting/sitting and 10:44 h recumbent; they underwent an average of 37 transitions per day between the respective posture and activity states (Troy et al. 2007).

Locomotion studies of seven emus (*D. novaehollandiae*) showed that they used a larger protraction angle than the ostrich to create a similar relative step length to the cassowary (Abourachid and Renous 2000).

*8.3.3 Cassowary (*Casuarius *spp.)*

The Northern Cassowary is distributed and endemic to coastal swamp and rainforests of northern New Guinea. The diet consists mainly of fruits and small animals. In the breeding season, the polygamous female lays three to five green eggs on a well-camouflaged nest prepared by the male; thereafter, she leaves the nest and eggs to find another mate. The male raises the chicks alone for about 9 months. Due to loss of the ongoing habitat and over hunting in some areas, the Northern Cassowary is evaluated as vulnerable on the IUCN Red List of Threatened Species. The major factors contributing to decline of the cassowary include: habitat

Fig. 8.5 A single cassowary in a enclosure at the Koala Centre, Brisbane, Australia

loss, fragmentation and modification, traffic accidents, visitor impacts, dogs, competition and nest predation by pigs, catastrophic events and disease.

Locomotion studies on this extremely aggressive bird are obviously lacking in the wild, although it could be feasible to place the bird in a large hold in a treadmill (Fig. 8.5). Abourachid and Renous (2000) filmed the locomotion characteristics of two Southern Cassowaries (*Casuarius casuarius*). The cassowary's speed is associated with its defensive need to use its sharp claws as daggers while leaping up into the air with deadly accuracy. The cassowary uses a larger protraction angle than the rhea to create a similar relative step length as in the emu (Abourachid and Renous 2000).

*8.3.4 Rhea: Greater Rhea (***Rhea americana***) and Lesser Rhea (***Rhea pennata***)*

Male rheas are very territorial during the breeding season. Persecution from farmers and egg gathering has led to a sharp population decline. They have little preference for cereals and grasses, but like cabbage. The rhea is found in grasslands, savannah, scrub forests, chaparral and desert areas, but is known to prefer areas with at least some tall vegetation. During the breeding season (which ranges from August to January in South America, April to August in North America), it stays near water. It is endemic to Argentina (Fig. 8.6), Bolivia, Brazil, Paraguay and Uruguay.

Locomotion studies of two Greater Rheas (*R. americana*) demonstrated that their given relative step lengths were similar to the ostriches, although the retraction

Fig. 8.6 An adult wild rhea in Argentina (courtesy of J.L. Navarro)

angle was smaller than the protraction angle in the rheas (Abourachid and Renous 2000). More locomotion studies are needed on wild rhea.

*8.3.5 Kiwi (*Apteryx *spp.)*

Captive management plans for kiwis produced by the New Zealand Department of Conservation were introduced to maintain and enhance the current abundance, distribution and genetic diversity. Captive management strategies include genetic studies, long-term captive populations, research (captive management, operation nest egg and adult breeding), collecting from wild populations and kiwis held abroad. The kiwi recovery plan (Robertson 2006) ascribed to the following goals: by 2006, there should have been successful encouragement and support of public and community protection of kiwi and their habitat, securing of representative populations of all wild kiwi taxa and identifying all genetically distinct kiwi populations (range, trends, threats and suitable management units) (Robertson 2006).

Using a treadmill with mirrors positioned vertically and laterally, a camera can record the locomotion patterns in an adult kiwi (Jones – personal communication). The bird uses its beak to position itself although a block is necessary to prevent the bird from falling back on the edge of the moving mat. Also, the experimenter needs to help the bird along. Considerably, more work is needed on kiwi movement in and outside their burrow.

Locomotion studies on the Brown Kiwi (*Apteryx australis*) showed that the bird increased its speed by increasing its stride length, principally via increase of the protraction angle (Abourachid and Renous 2000). Its hind limbs are constructed of a flexed jointed chain system, with its centre of mass being anterior. Its femurs are long and the knees act in yield, whereas the distal joints act in propulsion (Abourachid and Renous 2000). The kiwi also uses a larger protraction angle than the ostrich to obtain a longer step (Abourachid and Renous 2000).

8.3.6 *Tinamous (***Tinamus** *sp.,* **Crypturellus** *sp. and* **Rhynchotus** *sp.)*

The Red-winged Tinamou is ca. 40 cm in length. Its upper parts are brown barred with black and buff. The crown is mottled black, the throat is whitish, the fore neck and breast are cinnamon, and the remainder of the lower parts are grey-brown barred with black on the belly and flanks. The curved bill is horn-coloured with a blackish culmen. The adult has bright rufous primaries, which are mainly visible in flight. Juveniles are duller. It favours marshy grasslands and savannah. Its diet varies by season, taking insects and other small animals (even small mammals) in the summer, and switching to vegetable matter, such as fruits, shoots, tubers and bulbs, in the winter. Raising the tinamous in an environment similar to the chicken should ideally provide a good performance of growth rate, excellent carcass and breast yields, and perfect adaptation to meal and pelleted feeds.

There are few studies on the locomotion of tinamous. The walking parameters of the tinamous (*Tinamus solitarius*, *Crypturellus obsoletus* and *Rhynchotusrusfescens*) showed that they are paleognathous birds with a longer swing phase duration than neognathous birds, due to the former having a longer tarsometatarsus (Abourachid et al. 2005). The Tinamidae are therefore a sister group of ratites within the Palaeognathae taxon (Abourachid et al. 2005).

Preliminary locomotion studies on tinamous aged 3–4 years ($n = 20$, 10 hens and 10 cocks) did not show any significant differences (Anova analysis, $P < 0.05$) between body measurements between cocks and hens (de Queiroz and Cooper 2010). Girth was higher ($P < 0.05$) in cocks (35.80 ± 1.62 cm) than hens (33.80 ± 1.55 cm). In our sample, body weight was surprisingly higher in hens (786.60 ± 79.91 g) than cocks (747.20 ± 59.35 g), presumably due to either developing egg inclusion or greater age.

8.4 Further Running Performance Experiments

Using a method based on interactive B-spline solids for estimating and visualising biomechanically important parameters for animal body segments, validity was gained by reconstructing an ostrich body from a fleshed and defleshed carcass and comparing the estimated dimensions to empirically measured values from the original carcass (Hutchinson et al. 2007). Although the authors used the method to calculate the segmental masses, centres of mass and moments of inertia for an adult *Tyrannosaurus rex* (Hutchinson et al. 2007), it could be extended to compare the extinct ratite skeletons with the surviving ratites.

Time and space kinematic parameters of locomotion were compared in ratites (rhea, kiwi and Paleognatiforms) (Abourachid 2001). The results showed that in the two phases, stance and swing, the time and space parameters worked in opposite ways: the duration of the swing was constant, but its length increased with speed (Abourachid 2001). In all birds, a higher speed was achieved by a decrease of the stance duration, and an increase of the swing length (Abourachid 2001).

Similarly sized bipeds and quadrupeds use nearly the same amount of metabolic energy to run, despite dramatic differences in morphology and running mechanics (Roberts et al. 1998). Birds use on average 1.7 times more metabolic energy than quadrupeds. The cost of muscular force production determines the energy cost of running and suggest that bipedal runners use more energy for a given rate of force production because they require a greater volume of muscle to support their body weight (Roberts et al. 1998). This study can be extended to determine the metabolic cost of locomotion in ratites.

8.5 Relationships Between the Dynamics of Feeding and Locomotion

Kiwis are thought to detect their soil-dwelling invertebrate prey using their sense of smell (Cunningham et al. 2007). The unique position of the nares at the tip of the bill and the enlarged olfactory centres in the brain support this assumption (Cunningham et al. 2007). Kiwi possesses an arrangement of mechanoreceptors within pits suggesting that this sense may function in conjunction with, or be dominant over, olfaction during prey detection (Cunningham et al. 2007). This study could be extended to determine the olfactory role of the other surviving ratites.

Cranial kinesis in birds is induced by caudal muscles located on the cranium (Gussekloo and Bout 2005a). These forces are transferred into the moveable parts of the skull via the PPC. No clear bending zones are present in the upper bill, and bending is expected to occur over the whole length of the upper bill (Gussekloo and Bout 2005a). Muscle forces are more than sufficient to overcome bending forces and to elevate the upper bill. The resistance against bending by the bony elements alone is very low, which might indicate that bending of bony elements can occur during food handling when muscles are not used to stabilise the upper bill (Gussekloo and Bout 2005a). Model calculations suggest that the large *Processi basipterygoidei* play a role in stabilising the skull elements, when birds have to resist external opening forces on the upper bill as might occur during tearing leafs from plants (Gussekloo and Bout 2005a). *R. americana* showed that the feeding behaviour is a typical catch-and-throw behaviour, independent of the size of the food item (Gussekloo and Bout 2005b). Drinking is achieved by a scooping movement followed by a low-amplitude tip-up phase. During feeding, rhynchokinetic movements of the upper bill were observed. The specific morphology of the PPC is not the result of specific functional demands from palaeognathous feeding behaviour (Gussekloo and Bout 2005b). Again, further similar studies in the other ratites are important.

Frugivory and the dispersal curves produced by the southern cassowary, *C. casuarius*, showed that it consumed fleshy fruits of 238 species (Westcott et al. 2005). In feeding trials, seeds of 11 species were retained on average for 309 (mean) $\pm$ 256 (sd) min. Locomotion in forests would allow this bird to an estimated average dispersal distance of 239 $\pm$ 207 m (Westcott et al. 2005). Only 4% of seeds were dispersed further than 1,000 m. Observation of wild birds indicated that foraging and movement occur more frequently in earlier and later day (Westcott et al. 2005). This study should be extended to other ratites particularly whose environment is being threatened or disrupted by agriculture and/or wars.

In southwestern Australia, emus decrease their food intake when they start breeding in early winter and increase their intake during spring and summer when the breeding season and egg incubation are finished (Blache and Martin 1999). Long days increased food intake whereas short days decreased it. Day length seemed to affect appetite but not interest in food (Blache and Martin 1999). Determination of the distances covered by ratites in search of food during different day lengths would add useful knowledge to ecological considerations of ratite niches.

8.6 Conclusion and Recommendations

More funding might be devoted to exploratory work using a wireless technological real-time measurement or hold-conveyor belt measures of ratite physiological parameters using pervasive computing to measure precise changes in locomotion or movement via contraction of limb muscles and associated changes in heart rate, respiration, electromyography, blood–volume–pulse, electrocardiogram, respiration rate and electroencephalogram. The principles thereof could be adapted from that performed successfully in a human subject undergoing various activities (Cooper et al. 2008a, b, c). One downside, however, is the considerable expense of the apparatus. Using a similar objective in ratites, more studies of behavioural adaptation to terrestrial locomotion (Abourachid 2000) of the ratites is needed. Many more studies are needed to complete the aspects of ratite locomotion in the wild during various activities (foraging, reproductive displays, running, defence and sleeping). More extensive research is needed on the various stages of embryonic limb development in ratites to determine whether a developmental, genetic or incubational influence is predisposing weakening of the osteoblastic and osteoclastic associations. Leg tumours need more research. Given the global threats to wild ratite habitat, more studies need to be performed to determine the precise nature of their niches (wild, ranch or farming) to reduce the real threat of complete extinction.

References

Abourachid A (2000) Bipedal locomotion in birds: the importance of functional parameters in terrestrial adaptation in Anatidae. Can J Zoo 78: 1994–1998

Abourachid A (2001) Kinematic parameters of terrestrial locomotion in cursorial (ratites), swimming (ducks), and striding birds (quail and guinea fowl). Comp Bio Phy Part A: Mol Int Phy 131(1): 113–119

Abourachid A, Renous S (2000) Bipedal locomotion in ratites (Paleognatiform): examples of cursorial birds. Ibis 142: 538–549

Abourachid A, Höfling E, Renous S (2005) Walking Kinematics Parameters in some Paleognathous and Neognathous Neotropical birds. Orn Neo 16: 471–479

Blache D, Martin GB (1999) Day length affects feeding behaviour and food intake in adult male emus (*Dromaius novaehollandiae*). Br Poult Sci 40(5): 573–578

Bond WJ, Silander JA (2007) Springs and wire plants: anachronistic defences against Madagascar's extinct elephant birds. Proc Biol Sci 274(1621): 1985–1992

Charuta A, Majchrzak T, Czerwiński E, Cooper RG (2008) Spongious matrix of the tibio-tarsal bone of ostriches (Struthio camelus) – a digital analysis. Bull Vet Inst Pulawy 52:285–289

Cooper RG (1999) A discussion on ostrich chicks. Ost News 3(1): 3–9

Cooper RG (2000) Management of ostrich (Struthio camelus) chicks. World Poult Sci J 56: 33–44

Cooper RG (2002) Consideration on rearing and handling of ostrich chicks. In: Proceedings of the World Ostrich Congress, Warsaw, Poland, 26–29th September, pp 59–63

Cooper RG (2007a) Differences in stride between healthy ostriches (Struthio camelus) and those affected by tibiotarsal rotation. J S Afr Vet Assoc 78(1): 52–53

Cooper RG (2007a) Spread bow leg syndrome and associated pathology in ostrich (Struthio camelus) chicks aged 2–12 weeks. In: Proceedings of the XIV World Ostrich Congress, Riga, Latvia, 19–20th October, pp 34–39

Cooper RG (2008) Tennett A: locomotion comparison between the extinct terror bird and the living ostrich. Vet Hist 14(4): 371–375

Cooper RG, Horbańczuk JO (2004) Ostrich nutrition: a review from a Zimbabwean perspective. Revue Scientifique et Technique – Office International des Epizooties 23(3): 1033–1042

Cooper RG, Tennett A (2008a) Geometric limb similarity between two flightless birds: an extinct terror bird (Phorusrhacinae gen.) vs. the ostrich (Struthio camelus). In: Proceedings of the 4th International Ratite Science Symposium, Brisbane, Australia, 29th June–4th July, pp 17–18.

Cooper RG, Al-Muhtadi J, Ashford R (2008) Smart spaces with real-time physiological measurements and mitigation of stress. In: Proceedings of the 3 rd International Conference Pervasive Computing and Applications, Alexandria, Egypt, 6–8th October 2008, pp 3–8

Cooper RG, Mahrose Kh M, El-Shafei M (2008b) Spread bow leg syndrome in ostrich (Struthio camelus) chicks aged 2 to 12 weeks. Br Poult Sci 49(1): 1–6

Cooper RG, Naranowicz H, Maliszewska E, Tennett A, Horbańczuk JO (2008c) Sex-based comparison of limb segmentation in healthy and tibiotarsal rotation ostriches aged 14 months. J S Afr Vet Assoc 79(3): 142–144

Cunningham S, Castro I, Alley M (2007) A new prey-detection mechanism for kiwi (Apteryx spp.) suggests convergent evolution between paleognathous and neognathous birds. J Anat 211(4): 493–502

de Queiroz SA, Cooper RG (2010) Gender-based differences in stride and limb dimensions between healthy red-wing tinamou (*Rhynchotus rufescens*) Temminck, 1815. Turkish Journal of Zoology 35(1): 103–108.

Dzemski G (2007) Christian A: flexibility along the neck of the ostrich (Struthio camelus) and consequences for the reconstruction of dinosaurs with extreme neck length. J Morph 268(8): 701–714

Goetz JE, Derrick TR, Pedersen DR, Robinson DA, Conzemius MG, Baer TE, Brown TD (2008) Hip joint contact force in the emu (*Dromaius novaehollandiae*) during normal level walking. J Biomech 41(4): 770–778

Gussekloo SW, Bout RG (2005a) Cranial kinesis in palaeognathous birds. J Exp Biol 208(Pt 17): 3409–3419

Gussekloo SW, Bout RG (2005b) The kinematics of feeding and drinking in palaeognathous birds in relation to cranial morphology. J Exp Biol 208(Pt 17): 3395–3407

Gussekloo SW, Janssen BA, George Vosselman M, Bout RG (2000) A single camera roentgen stereophotogrammetry method for static displacement analysis. J Biomech 33(6): 759–763

Gussekloo SW, Vosselman MG, Bout RG (2001) Three-dimensional kinematics of skeletal elements in avian prokinetic and rhynchokinetic skulls determined by Roentgen stereophotogrammetry. J Exp Biol 204(Pt 10): 1735–1744

Hallam MG (1992) The Topaz Introduction to Practical Ostrich Farming. Superior Print and Packaging, Harare, Zimbabwe, 160

Harshman J, Braun EL, Braun MJ, Huddleston CJ, Bowie RC, Chojnowski JL, Hackett SJ, Han KL, Kimball RT, Marks BD, Miglia KJ, Moore WS, Reddy S, Sheldon FH, Steadman DW, Steppan SJ, Witt CC, Yuri T (2008) Phylogenomic evidence for multiple losses of flight in ratite birds. In: Proceedings of the National Academy of Sciences USA 105(36): pp 13462–13467

Horbańczuk JO (2002) The Ostrich. Warsaw, Poland, 190

Hutchinson JR, Ng-Thow-Hing V, Anderson FC (2007) A 3D interactive method for estimating body segmental parameters in animals: application to the turning and running performance of Tyrannosaurus rex. J Theor Biol 246(4): 660–680

Jindrich DL, Smith NC, Jespers K, Wilson AM (2007) Mechanics of cutting maneuvers by ostriches (Struthio camelus). J Exp Biol 210(Pt 8): 1378–1390

Madeiros C (2007) Essential information for the ratite veterinary surgeon. In: Proceedings of the XIV World Ratite Congress, Athens, Greece, 13–14th October 2007, pp 28–45

Maxwell EE, Larsson HC (2007) Osteology and myology of the wing of the Emu (*Dromaius novaehollandiae*), and its bearing on the evolution of vestigial structures. J Morph 268(5): 423–441

Palmieri G, Sanna M, Bo Minelli L, Botti M, Gazza F, Di Summa A, Santamaria N, Passantino L, Maxia M, Acone F (2003) On the sensitive innervation of the ostrich's foot pads. Ital J Anat Emb 108(1): 25–37

Patak AE, Baldwin J (1998) Pelvic limb musculature in the emu *Dromaius novaehollandiae* (Aves: Struthioniformes: Dromaiidae): adaptations to high-speed running. J Morph 238(1): 23–37

Roberts TJ, Kram R, Weyand PG, Taylor CR (1998) Energetics of bipedal running I. Metabolic cost of generating force. J Exp Biol 201(Pt 19): 2745–2751

Robertson HA (2006) Kiwi (Apteryx spp.) recovery plan. Biodiversity Recovery Unit. Department of Conservation, Wellington, New Zealand, p 24

Rubenson J, Heliams DB, Lloyd DG (2004) Fournier PA: Gait selection in the ostrich: mechanical and metabolic characteristics of walking and running with and without an aerial phase. Proc Biol Sci 271(1543): 1091–1099

Rubenson J, Lloyd DG, Besier TF, Heliams DB, Fournier PA (2007) Running in ostriches (Struthio camelus): three-dimensional joint axes alignment and joint kinematics. J Exp Biol 210(Pt 14): 2548–2562

Samorek-Salamonowicz E, Kozdruń W, Czekaj H, Kozanecki P (2002) Case of the lameness in ostrich. In: Proceedings of the World Ostrich Congress, Warsaw, Poland, 26–29th September 2002; 250

Schaller NU, Herkner B, Prinzinger R (2005) Locomotor characteristics of the ostrich (Struthio camelus). I: morphometric and morphological analyses. In: Proceedings of the 3 rd International Ratite Science Symposium and XII World Ostrich Congress, 14–16th October 2005; pp 83–90

Smith NC, Wilson AM, Jespers KJ, Payne RC (2006) Muscle architecture and functional anatomy of the pelvic limb of the ostrich (Struthio camelus). J Anat 209(6): 765–779

Smith NC, Payne RC, Jespers KJ, Wilson AM (2007) Muscle moment arms of pelvic limb muscles of the ostrich (Struthio camelus). J Anat 211(3): 313–324

Troy KL, Lundberg HJ, Conzemius MG, Brown TD (2007) Habitual hip joint activity level of the penned EMU (Dromaius novaehollandie). Iowa Ortho J 27: 17–23

Westcott DA, Bentrupperbäumer J, Bradford MG, McKeown A (2005) Incorporating patterns of disperser behaviour into models of seed dispersal and its effects on estimated dispersal curves. Oecologia 146(1): 57–67

Wings O, Sander PM (2007) No gastric mill in sauropod dinosaurs: new evidence from analysis of gastrolith mass and function in ostriches. Proc Biol Sci 274(1610): 635–640

Wöhr AC, Schulz A, Erhard MH (2005) Animal welfare aspects regarding the raising of breeding ostriches in Germany. Deutsche Tierarztliche Wochenschrift 112(3): 87–91 [in German]

Chapter 9
Ratite Health: Welfare Implications

D. Black and P.C. Glatz

Abstract The Codes of Practice developed for ratites in a number of countries place considerable importance on ensuring that the health of birds is closely monitored and that appropriate vaccines are used to prevent disease. Likewise, if the health of ratites is compromised, there are clear guidelines on actions required by farmers to improve the health of birds. The action required by persons in caring for the health of birds is a significant welfare responsibility. The welfare of ratites when using health as an indicator can range from minor ailments or infections with minimal concerns to the major health issues that cause considerable pain and discomfort to animals. When the death of the animal is potentially involved, immediate action is necessary to prevent mortality. Effective implementation of preventative health programmes that involve record keeping, preventative health measures, quarantine principles, biosecurity practices and monitoring of management practices can improve the efficiency of ratite farms and also improve bird health and welfare.

Keywords Codes of Practice · Disease · Health · Inspection · Parasites · Vaccination · Welfare

9.1 Introduction

The introduction of Codes of Practice for Ratites in a number of countries (e.g. Standing Committee on Agriculture and Resource Management 2003; Animal Welfare Advisory Committee 1998) has meant that persons who are responsible

D. Black (✉)
ASAEL Consultancy Services, RMB 2235, Moama NSW 2731, Australia
e-mail: doug@microchips.com.au

P.C. Glatz
SARDI, Roseworthy Campus, University of Adelaide, Roseworthy, SA 5371, Australia
e-mail: phil.glatz@sa.gov.au

P.C. Glatz et al. (eds.), *The Welfare of Farmed Ratites*, Animal Welfare 11,
DOI 10.1007/978-3-642-19297-5_9,

for ratites must ensure that the bird's health and welfare are maintained. The basic requirements of the Codes of Practice (e.g. in Australia and New Zealand) that relate to health include provision of sufficient food and water to sustain health, protection from disease (including those diseases that are caused by poor management) and avoidance of pain, distress, suffering and injury in birds.

Good management is essential to maintain ratite welfare, including taking action to minimise contact of the ratites with other bird species (wild or farmed) and other animals. Appropriate hygiene, proper housing, and brooding and appropriate stocking density are essential when the welfare of ratites is being judged. For example, high stocking rates for yearling and adult groups of ostriches and emus could result in poor bird welfare as a result of aggressive behaviour and injuries during the breeding season (Glatz 2001). Newly acquired stock need to be quarantined from the existing stock for 4–6 weeks to minimise risk of the introduction of a disease (Glatz 2001). The housing facilities and equipment used in ratite farming need to be cleaned and disinfected as often as is practicable before restocking to prevent the carry-over of the disease-causing organisms to incoming birds. Ratites should not be kept on land that has become contaminated with organisms that cause or carry disease to an extent as it could seriously prejudice the health of ratites (Standing Committee on Agriculture and Resource Management 2003; Animal Welfare Advisory Committee 1998).

Good management requires that sick and injured ratites need to be treated without delay and isolated if necessary. Records of sick animals, deaths, treatment given and response to treatment need to be kept to assist disease investigations. Ratites that have an incurable disease, irreparable injury or painful deformity that create unacceptable levels of suffering should be humanely euthanased. Preventative health programmes and performance targeting can greatly contribute to the efficiency and ultimate viability of ratite farming (Glatz 2001).

This chapter identifies the major health issues in ratites, the extent to which welfare is compromised on the basis of pain and discomfort and approaches that should be taken to improve ratite flock health.

9.2 The Importance of Health as an Animal Welfare Issue

The main health issues in ratites are disease, injury, functional impairment and pain, which all impact on the welfare of birds. The degree to which these health issues impact on the welfare of birds is based on birds suffering long periods of stress, pain, debilitation and suffering.

The intensification of ratite farming has resulted in an improvement in ratite nutrition, housing and health. However, most ratite species evolved in environments where they could roam over large areas. In the confined environments, ratites have some difficulty with being housed in small areas, adapting to reduced feeding space and higher stocking densities. This has led to some health issues, particularly leg problems and injuries of birds running into fences and rapid spread of disease (see Chaps. 4 and 9). Also, in many cases, the feeding of small volumes of high caloric density and low fibre content feeds compared with the natural diet has often led to serious digestive disorders (Black 1998d).

Some animal welfare groups consider that confinement of ratites does not allow birds' complete freedom of movement and that the farm environment is totally different from the nomadic system in which ratites have evolved (European Communities 1998). The housing of birds under intensive farming systems has meant that the health challenges on birds require more careful management. The risks to health and mortality increase as ratites are moved into intensive farming systems (Glatz 2001).

Studies of ostrich chick mortality indicate that most deaths occur before 2 months of age, and the chicks of this age are most likely to die of infectious diseases. From 2 months of age, death is usually the result of leg problems, trauma or proventricular impactions. Mortality rates from all causes drop significantly after 4 months. In South Africa, chick mortality is reported to be as high as 40–50% up to 3 months and 10–30% from 3 to 6 months. Mortality from 3 to 12 months dropped to 5% (Verwoerd et al. 1999). Therefore, problems with chicks are a major welfare issue in ratite production (Chap. 2).

9.3 Health and Welfare Responsibilities of Ratite Stockpersons

The person caring for ratites is responsible for their welfare. Evidence of behavioural changes in ratites can be a good indicator of poor health or distress (see Chaps. 5 and 6). Symptoms include birds that are alone, standing hunched up, in sternal recumbency, lethargic, not eating, loss of vigour, changes in faeces or urine, coughing, abnormally increased rate and effort to breath, lameness, dull and brittle feathers and swelling on the body or legs (Glatz 2001). Preventing infectious disease and internal and external parasites is also an important strategy to maintain bird welfare and health at optimum levels. Sick and injured ratites need to be treated without delay and isolated if necessary. Prompt attention to health issues in ratites means that the welfare of birds is not being compromised and avoids pain or discomfort or stress.

9.3.1 Farm Management

Keeping good records is an important part of high quality farm management (Barnett et al. 2001).

Adequate records will assist in the detection of any health problems. Accurate information should be kept on the source of all stock. Records of sick animals, deaths, treatment given and response to treatment should be maintained to assist disease investigations. Dead stock should be promptly removed and, if not required for post-mortem examination, should be disposed of in a hygienic manner such as by deep burial or incineration. Ratites with an incurable disease, irreparable injury or painful deformity that cause unacceptable levels of suffering should be humanely killed. While disease outbreaks can often involve primary or virulent pathogens such as Pseudomonas, Salmonella and virulent *Escherichia coli* strains, Newcastle Disease virus or Avian Influenza virus, the greatest single contributing factor to disease outbreaks on ostrich farms is poor management (Black 2001).

9.3.2 Inspections of Birds

Regular inspection of the birds is considered to be critical for the welfare of the entire flock. It is good farming practice to undertake inspection as a separate and distinct procedure as opposed to casual observation while other tasks are performed. The frequency and level of inspection should be related to the needs of the ratites. Under certain circumstances, more frequent inspections may be required, such as during hot weather, during outbreaks of disease or when groups of birds have been mixed (Glatz 2001). Injured birds should be removed for treatment without delay. In Australia, compliance with the intent of the Code of Practice for Ratites in relation to dead, sick or deformed birds can only be achieved by frequent inspections. Similarly, without frequent inspections of feeders and drinkers and other equipment such as holding yards, the welfare of the birds will be compromised.

There are two levels of inspection required on ratite farms; those by stockpersons and those undertaken by veterinarians. For stockpersons, the checking of birds is considered as one of the most important daily management tasks. Regular inspection is essential in the first 48 h after hatching and is considered as one of the tasks critical in protecting the subsequent production and welfare of the birds. Regular inspection may prevent issues such as heat stress, cold stress and injuries, which may have a long-term impact on performance and welfare. While the Code of Practice recommends at least one inspection per day, industry would generally exceed this. While it is difficult to provide well-defined inspection frequencies, it is expected that during the first week, there would be a minimum of four inspections daily for birds in the first week. It is recommended that there should be a 1–2 hourly inspection for hatchlings, and the frequency could be reduced as the age increases to once or twice daily for adults (Glatz 2001). These inspections should be spread out during daylight hours and the frequency should be increased if a problem is identified.

It is expected that veterinarians would inspect flocks when farmers report higher than normal levels of bird and embryo mortality or if birds are stressed, in pain and suffering. The level of mortality that prompts farmers to request veterinary inspection can be included in an On-Farm Surveillance Plan. A well-designed and carefully implemented flock health scheme is a feasible way of achieving improved performance, welfare and profitability through focussing on improvement of these areas (Black 2001; Doneley 1996b). Investigation of hens for problems with high rates of embryo mortality or egg infection can involve clinical examination of the hen, including oviduct swabbing and culture/sensitivity testing and sometimes antibiotic treatment.

An accurate diagnosis and timely implementation of appropriate treatment and control measures are vital to control the spread of disease on a large-scale ratite farming operation. Investigation of problems in the incubation and hatching facility, thorough examination of fertile eggs that have failed to hatch, clinical examination of sick birds and the autopsy examination of dead birds are all important measures (Black 1998c). Thorough autopsy examination by an experienced veterinarian of any deaths in significant numbers or in unexplained circumstances is critical and a significant welfare responsibility for ratite farmers. As in other avian species, many diseases in ratites can appear clinically similar and histopathological, bacteriological and sometimes, virological testing may be needed to establish a diagnosis. There is a need to look beyond the pathological or clinical diagnosis to identify any possible predisposing management factors that may have initiated or perpetuated the disease (Black 2001).

9.3.3 Monitoring Flock Health

Monitoring flock health via regular blood samples for serological testing is also considered to be an important component of sustaining bird health and welfare. Serological testing throughout the life of the flock is a way of monitoring flock health and effectiveness of vaccinations. Several veins are readily accessible for venipuncture in ratites. The right jugular vein is suitable for younger birds and emus. Care should be taken while using this vein as the thin-walled jugular can tear easily, leading to the formation of a potentially fatal haematoma (Black 1995). The basilic vein can be used in ostriches, but the small vestigial wings in other ratites such as emus make this vein unsuitable. The medial metatarsal vein is present in all species, but the thinner skin and the unique behavioural response to "hooding" of the ostrich make this the only species where use of this vein is feasible. Care must be taken that adequate restraint is in place before this vein is utilised. A blood smear should be made immediately, and the rest of the sample placed in plain tubes for serological testing or, preferably, lithium heparin for biochemistry examination. If the samples cannot be analysed rapidly, it is advisable to centrifuge the sample and refrigerate the plasma or serum as the value of the information depends largely on the quality of the serum/plasma sample. The determination of serum antibody titres in a flock can be important in detecting exposure to certain disease agents and the monitoring of vaccination programmes. Haematology and biochemistry examination can aid in the diagnosis of disease. When investigating a possible disease problem, birds showing suspect or clinical signs should be used. The number of samples required for accurate disease surveillance serology will depend on the size of the flock.

To investigate a reproductive disease problem such as infertility, embryonic deaths, poor shell quality, irregular laying and misshapen eggs, the reproductive records need to be carefully analysed and the breeding birds in question need to be observed and physically examined. For some of these problems relating to the hen, a swab can be taken of the distal oviduct for cytology and culture. With the hen hooded and minimally restrained, a gloved and lubricated hand is placed in the urodeum and the oviduct opening. This opening has a mild sphincter and is found in the 10 o'clock position of the cranial wall of the urodeum. A finger can be passed through this opening into the distal oviduct, and then a guarded mare swab introduced alongside the finger. A limitation of this technique is that only the caudal 10–15 cm of the oviduct can be accessed – a disease process occurring more proximally may be missed. Common pathogens isolated include *E. coli*, *Pseudomonas* spp. and *Klebsiella* spp. Faecal or urine/urate contamination is not uncommon (Black 1995, 1998b).

9.3.4 Physical Examination

Some birds suspected of clinical disease need to be caught and physically examined. An important component of bird inspection is to safely restrain the bird to enable this physical examination to occur. A systematic and thorough evaluation is invaluable. For example, a detailed guide to the examination of the ostrich is provided in Black (1995). A summary of his technique can be applied to all ratites. First, the bird is assessed for its gait, respiration and feathering. Next, the bird is caught, identified (either by leg band, neck tag or microchip etc) and the following inspections are performed:

- Externally examine the eyes, beak, nares, ears and oral cavity
- Check the body condition by palpating the spine
- Auscultate the thorax, determine the heart and respiration rate and check for abnormal respiratory and cardiac sounds at several sites
- Auscultate the proventriculus and ventriculus
- Palpate the abdomen, including the proventriculus and ventriculus
- Observe the feet and legs for abnormalities
- Palpate the limbs and determine the alignment of the tibiotarsus
- Examine the skin and feathers for abnormalities and parasites
- Examine the vent and digitally sex the bird if necessary

A standard routine allows accurate recording and ensures no part of the examination is overlooked.

9.3.5 Culling

Culling of sick ratites may be required to reduce the potential for transfer of disease and to reduce any pain and suffering of birds and is therefore an important welfare issue for ratites. Although it is unknown whether unthrifty cull birds are in pain, it is ethical to assume that animals feel pain and distress in a similar manner to humans and thus, if they cannot be successfully treated, they should be culled as quickly as possible. Timely culling is considered to be critical for the welfare of the entire flock of birds (Glatz 2001). From a welfare perspective, it is important to identify birds to be culled and to humanely kill them. On farms with large numbers of breeders, there will always be a significant number of poor or non-performing breeding birds. Problem hens may be identified using ultrasound examination especially at the end and beginning of each breeding season. This technique can identify conditions such as ovarian inactivity, egg retention, egg yolk peritonitis and abdominal cancer.

Aggressive or disruptive breeding birds (especially males) can usually be identified via observation by personnel on the farm. Stress due to harassment by these birds can lead to lowered egg production and trauma in the hens of that particular breeding unit. They can also cause lowered fertility rates and traumatic injuries in other male birds due to an increase in fighting behaviour. These birds are a menace and a danger to personnel, especially those involved in egg collection. The provision of shelter and "safety zones" in breeder paddocks will minimise the negative effect of these males (see Chap. 2). Ideally, the worst offenders should be culled or tried in other units to see whether the problem behaviour disappears and, if not, should then be definitely culled (Black 1995).

9.3.6 Autopsy Examinations

Timely autopsy examinations are considered good farming practice to be able to identify problems and to provide an opportunity to minimise such problems in the future and maintain optimum welfare for birds. With experience, farmers can conduct their own autopsies for some common causes of death and also identify situations in which expert advice is required. The major elements are firstly to spend time with the farm veterinarian while he or she is conducting routine autopsies to observe techniques and understand normal from abnormal. If the cause of death is not obvious to the farmer or samples need to be collected then the experienced veterinarian will need to be involved. Recording mortalities, including culled birds, is considered to be critical for the welfare of the entire flock of birds (Barnett et al. 2001).

9.3.7 Disposal of Dead Birds

The timely removal of dead/culled birds and disposal of dead birds is an important health and biosecurity issue to prevent the build-up of pathogenic micro-organisms and the transmission of disease to healthy birds. Common methods for disposal are burial on site in properly covered burial pits (solid roof and a tightly fitting access cover) and using incineration. It is important that compost and burial sites are covered to prevent scavenging by birds and rodents. Open trenches should not be used (Glatz 2001).

9.3.8 Vaccination

In some countries, the use of vaccines to prevent infectious disease is considered to be one of the important welfare responsibilities of farm managers. Vaccinations and other treatments of birds should be undertaken by people skilled in the procedures. Medication should only be used in accordance with the manufacturer's instructions or otherwise advised by the experienced veterinarian (Barnett et al. 2001).

Vaccination is defined as the introduction of an attenuated or inactivated disease agent into an animal to produce a degree of protection against a disease through an immune response. It is critical that the immune system is able to recognise antigens (foreign proteins), produce antibodies and develop a memory of the antigen for a subsequent response. This latter response is enhanced by exposure to a specific microorganism, either by natural exposure at the farm or re-vaccination. Although vaccination has long been the preferred method of disease control, many other factors also play an important role in the overall success of a vaccination programme. These factors include management, feed quality, environmental conditions, field exposure, flock health, biosecurity, genetics, age and route of vaccine administration (Bourke et al. 2003).

Poor administration is the most common cause of vaccine failure in ratites. Planning, training and attention to detail will result in better vaccination methods and improve disease control (Bourke et al. 2003).

Live vaccines are susceptible to a number of agents including disinfectants, chlorinated water and other water sanitising materials commonly used on farms. Basic steps must be taken to avoid exposing the vaccine to these chemicals. These include washing hands (ensuring the soap is thoroughly rinsed off), only mixing vaccine in a clean area and on a clean surface to ensure that surfaces that carry traces of disinfectant do not contaminate the vaccine during mixing with water.

9.3.9 Specific Vaccinations

The most common pathogens targeted for vaccination include *Clostridium perfringens*, Poxvirus, Newcastle Disease and autovaccines of bacterial isolates. Vaccination against Avian Influenza and Anthrax may also be used. These pathogens can cause disease and death and the birds may suffer considerable stress, pain and possible dehydration and starvation prior to death if subjected to the full disease course. Thus, such diseases can result in very poor health and welfare and vaccination is required to help protect the health and welfare of the bird.

For example, *C. perfringens* enterotoxaemia and enteritis usually involves deaths of chicks, juveniles and young adult ostriches (Huchzermeyer 1998) and is often related to inadequate fibre intake, transport and relocation, excessive lucerne intake or sudden or inappropriate changes in diet (Huchzermeyer 1994a, b, c). For this reason, it is critical that if a *C. perfringens* vaccination is available and protective against all potentially pathogenic *C. perfringens* types that are prevalent on ostrich farms, then it should be employed to minimise the welfare impact of this disease on ratites.

Poxviral outbreaks (caused by the fowl poxvirus) have also been observed in ostrich chicks in a number of countries. Scabby lesions of the beak and the non-feathered skin around the eyes and eyelids are the most common symptoms. In some cases, granulomatous lesions can lead to depression, lack of vision, reduced food intake and some mortality. It is important to consider the welfare of the bird, particularly the discomfort associated with the sometimes spectacular eyelid lesions, and implement control measures that include topical and systemic treatment of secondary lesions, vector control and vaccination using a commercial fowl pox vaccine to prevent further outbreaks (Raidal et al. 1996).

Avian influenza is a bird disease caused by a number of strains (or isolates) of serotypes (or subtypes) of the influenza A virus that are endemic in birds. Influenza A virus is the only species in the genus influenza virus A of the orthomyxoviridae family of viruses (http://en.wikipedia.org/wiki/Avian_influenza). The influenza viruses occur naturally among birds. Wild birds can carry the virus but may not suffer clinical disease and so can be a source of transmission. Domesticated birds may become infected with avian influenza virus through direct contact with infected waterfowl or other infected poultry, or through contact with surfaces or materials (such as water or feed) that have been contaminated with the virus. The high extremes of virulence infection with avian influenza viruses may cause disease that affects multiple internal organs and has a mortality rate that can reach 90–100% often within 48 hours (http://www.cdc.gov/flu/avian/gen-info/facts.htm). Therefore, Avian Influenza represents one of the most potent diseases and has a major impact on bird welfare. Avian Influenza outbreaks in ostriches have been reported in South Africa (Allwright 1996; Huchzermeyer 1998), primarily involving symptoms of respiratory disease and ocular discharge. The severity of the symptoms is worse in young chicks less than 4 months of age and especially with concurrent bacterial or fungal infections. Hence, farmers need to be on alert to ensure that the health and welfare of the birds are protected. An inactivated autogenous emulsified vaccine had been used in South Africa where it controlled morbidity and mortality, but failed to prevent shedding of the virus. The elimination or reduction of contact between ostriches and wild birds suspected as being carriers of the disease is the most important measure in preventing recurrences of the disease.

Newcastle disease (ND) is another disease that has a critical impact on welfare and health of ostriches. It was found in zoo birds in the 1950s, in commercial ostriches in Israel in 1989 and in southern Africa during the 1990s (Alexander et al. 1998). Both live and inactivated ND poultry vaccines are protective in ostriches, but are usually given more frequently and at much higher doses than recommended for poultry (Alexander et al. 1998). Clinical signs of ND consist of general depression and symptoms associated with the central nervous system such as recumbency, paralysis or obvious neck weakness.

ND vaccination is required when the disease is endemic. Horizontal spread is mainly via the drinking water and faecal–oral transmission (Verwoerd 1998). The outbreaks usually only involve a few birds, but mortality of 30% can be observed (Huchzermeyer 1998). Outbreaks are generally well controlled through measures to limit horizontal transmission coupled with vaccination. Vaccination schedules will vary according to the region, the degree of risk of the disease and the antibody status of the breeding flock and chicks (Blignaut et al. 1998; Schaetz et al. 1998). The use of ND vaccines is often regulated in many countries and sometimes prohibited.

The methods for ND vaccination of ostriches recommended by Huchzermeyer (1994b) to protect their health and welfare are as follows: (1) primary live vaccine by eye drop, (2) inactivated oil emulsion (twice the recommended chicken dose) 3 weeks later and (3) inactivated oil emulsion (twice the recommended chicken dose) every 6 months in growing birds and every year in breeding birds. However, Madeiros (1997) recommended different methods for birds with and without maternal immunity, which should be used since maternal antibodies affect the effectiveness of vaccination. The methods he recommended were (doses relative to the poultry dose) for those without maternal antibodies: (a) 2 weeks old, live vaccine (×5) eye drop plus inactivated oil emulsion vaccine (×1); (b) 1 month old, inactivated (×2); (c) 2 months old, inactivated (×6); (d) 6 months old, inactivated (×6); (e) 12 months old, inactivated (×10) and (f) annually, inactivated (×10). For those with maternal antibodies: (a) 45 days old, inactivated oil emulsion (×6); (b) 70 days old, inactivated oil emulsion (×6); (c) 6 months old, inactivated oil emulsion (×6); (d) 12 months old, inactivated oil emulsion (×10) and (e) annually, inactivated oil emulsion (×10).

Autogenous vaccines have been used to control some cases of ongoing bacterial infections, when treatment with the appropriate use of antibiotics has failed to adequately control the disease (Doneley 2006). Autogenous bacterial vaccines have often proven only partially effective in adequately controlling disease and preventing further outbreaks (Black 2001). Anthrax spore vaccines have also been used to protect ostriches in endemic anthrax areas.

9.3.10 Prophylactic Antibiotic Treatment

Prophylactic antibiotic treatment is only advisable as a temporary preventative measure in acute situations. The primary focus should always be to identify and correct the inevitable management problems that are leading to the emergence of the disease and subsequent health and welfare problems of birds. It should be remembered that in many countries ratites are classified as food-producing animals and a written drug withholding period advice is mandatory and the use of some antibiotics may be prohibited. An example of prophylactic antibiotic treatment would be the inclusion of zinc bacitracin in food to help to prevent Clostridial disease when an effective vaccine is unavailable.

9.3.11 Internal Parasite Control

Parasites are a health and welfare problem for ratites on some farms. Sotiraki et al. (2001) took faecal samples from farms in Greece to examine the gastrointestinal parasites in ostriches and found 90% of the ostrich farms were infected by parasites; the most common being protozoa.

The two significant internal parasites of ostriches are *Libyostrongylus douglassi* (wire worm) and *Houttuynia struthionis* (tape worm). Libyostrongylus infestations can cause significant problems and mortalities in ostrich chicks due to anaemia and inflammation of the proventriculus and gizzard, and therefore, protection against these parasites is a high priority welfare issue, although some countries have not confirmed their presence (Barton and Seward 1993; Button et al. 1993). Wireworm causes a compensatory production of thick mucous that affects digestion (Craig and Diamond 1996). Mukaratirwa et al. (2004) recommended that regular de-worming of birds can reduce the contamination of the ostrich wireworm. Prophylactic treatment programs will vary according to the infestation levels of the parasites, but because of the relative resistance of Libyostrongylus eggs, preventative measures should be based on good management and welfare by reducing stocking rates, the use of uncontaminated paddocks, removal of fresh faeces, quarantine and treatment of any newly introduced birds and regular monitoring through larval culture of faecal samples from paddocks (Black 1995). Another roundworm, *Codiostomum struthionis* is found in the colon of the ostrich, but is of little pathogenic significance. However, the eggs of this parasite closely resemble those of the pathogenic *L. douglassi* and hence larval culture is required for diagnosis.

The intestinal nematode *Deletrocephalus dimidiatus* in a flock of greater rheas was reported in the UK (Taylor et al. 2000). The clinical signs of infections included weak, diarrhoeic chicks with high flock mortality in heavy infections (Craig and Diamond 1996). Several anthelmintics may be of use in treating these nematode infections in rheas.

Levamisole is recommended for use in ostriches at an oral dose of 7.5 mg/kg, with activity particularly against the proventricular worm, *L. douglassi* (Huchzermeyer 1994a). Dose rates for benzimidazole anthelmintics suggested for use in ostriches (Anon 1998) include fenbendazole given orally at a dose rate of 15–45 mg/kg and oxfendazole given at 5 mg/kg. Ivermectin has also been used.

9.3.12 External Parasite Control

It is not uncommon to have heavy infestations of feather lice (*Struthiolipeurus struthionis*) and quill mites (*Gabucinia* spp.) on ostriches and rheas (Craig and Diamond

1996). These parasites cause relatively minor health and welfare issues for birds, but can cause damage to the feathers and possibly the skin of infected ostriches either directly by the parasites or indirectly through the excessive preening or rubbing of the skin, leading to diminished economic value of feathers and hide. Effective quarantine procedures can prevent the entry of these parasites to farms, but if there is a significant infestation, routine treatment is based on using systemic and/or topical parasiticide or insecticide treatments such as malathion spray (if approved in that country), pyrethrin-based spray, topical or systemic ivermectin or moxidectin.

Ticks can cause damage to the skin, resulting in poor quality leather. Paralysis due to tick bites is limited to the geographical areas where the offending tick occurs. Generally, chicks and young birds are most likely to be affected and many birds often recover if the ticks are removed. Craig and Diamond's (1996) recommended treatment for ticks consists of direct application of insecticides as mentioned earlier.

9.3.13 Control of Protozoal Parasitism

Infections with protozoa such as Hexamita-like flagellates, *Histomonas meleagridis*, *Giardia* spp., *Cryptosporidia*, *Microsporidia* and *Balantidium* have all been reported in ostriches. Although these infections are rarely pathogenic and possibly secondary to other pathogens, prophylactic treatment is sometimes required on farms where protozoal disease has been a problem. Histomoniasis (caused by the protozoan *H. meleagridis*) is a protozoal disease, which damages the caeca and liver (McDougald 1997). The clinical signs of histomoniasis in rheas include depression, anorexia and yellowish diarrhoea (McMillan and Zellen 1991), but no clinical signs of this disease have been reported in ostriches.

9.4 Major Diseases of Ratites

9.4.1 Diseases of the Integument

The welfare of birds afflicted with diseases of the integument rarely causes serious welfare issues that result in death except when the bird is unable to eat and drink.

Huchzermeyer (1998) reported mycotic dermatoses involving *Aspergillus* spp., *Trichophyton* spp. and *Microsporum* spp. Treatment with 10% enilconazole washes (Imaverol, Janssen) was the recommended treatment.

As stated previously, Avipoxvirus has been reported in ostrich chicks in South Africa, Israel, USA and Australia (Shane and Tully 1996; Raidal et al. 1996). The proximity of poultry and the presence of suitable vectors contribute to outbreaks. Chicks from the age of 2 weeks are affected, with both cutaneous and diphtheritic forms reported. Vaccination from 10 days of age with Fowl Pox vaccine, combined with biting insect control, will usually limit or prevent outbreaks and assist with improving the health and welfare of the bird.

Hyperkeratosis of the skin around the head, eyelids, beak commissures, neck and feet has been observed in ostrich chicks, and is thought to be nutritional in origin. Supplementation with zinc and biotin is often curative.

Skin lacerations are not uncommon in captive ratites from claws of other birds and with injuries inflicted by collisions with wire fences, a frequent occurrence on some farms (Glatz 2008). Although these lacerations respond well to treatment especially in the neck and upper body, it must be noted that scarring will significantly detract from the hide value. Fence construction and handling procedures must be done with the aim of minimising any such injuries and was included as a quality assurance check in a ratite skin audit (Glatz 2001). Lacerations of the lower limb can result in severe secondary bone problems from avascular necrosis due to damage to the periosteum.

Bacterial dermatitis and folliculitis does not appear to be common, but can occur in excessively humid climates. Early recognition and treatment is essential to prevent or minimise damage to the hide.

As stated earlier, ectoparasites are common in ratites, with ticks, lice and mites being found on most species. While the primary clinical focus is on the damage caused to plumage and hide by lice such as the ostrich louse (*S. struthionis*) and mites such as the quill mite (*Gabucinia bicaudatus*), other problems have been recorded. Tick paralysis, caused by the Ixodid tick *Hyalomma truncatum*, has been reported in South Africa, whereas the Argasid tick *Argas persicus* can cause anaemia in chicks and transmit aegyptianellosis from chickens to ostriches (Craig and Diamond 1996). Regular spraying with 2% malathion or pyrethrins needs to be part of the husbandry of ratites and to ensure their welfare is maintained.

9.4.2 Diseases of the Gastrointestinal Tract

Congenital deformities of the beak are occasionally seen in ostrich chicks. The more common of these deformities include scissor beak and a downward deviation of both rhinotheca and gnathoceca. These rarely cause any significant problems to the chick, and treatment is rarely warranted. Severe deformities can interfere with the prehension of food and unless the chick is valuable, euthanasia may be the preferred option. Manganese deficiency in parental diets is reported to cause a shortening of the lower beak (Huchzermeyer 1998).

The fungal infection candidiasis in ratite chicks can result from immunosuppression, malnutrition or prolonged antibiotic therapy and can become a significant welfare issue for birds. Plaque-like lesions can be observed in the oropharynx, although the infection can extend through to the ventriculus. Cytology and/or culture are diagnostic. Treatment with nystatin or ketoconazole is usually successful.

Other fungal infections, including zygomycosis and megabacteriosis (Avian Gastric Yeast) an infection caused by *Macrorhabdus ornithogaster*, have been reported in ostrich chicks (Shane and Tully 1996; Huchzermeyer 1998).

Contact with wild birds or commercial poultry, as well as environmental stresses including high stocking density and poor hygiene, lead to infectious disease in ratites (Shane and Tully 1996). Ratite chicks (3 weeks of age) are more susceptible to bacterial infections if the immune system is not yet well developed (Minnaar and Minnaar 1995). Most diseases are related to farm management including inappropriate feed and inadequate or contaminated water supply (Samson 1997), climate, stress, hygiene and incubator/brooder management. Proper management of these areas can reduce the risk of disease infection and improve bird welfare.

Details of bacterial (*Anthrax*, *Salmonellosis*, *E. coli*, *Colibacillosis*, *Pausterellosis* and *Tuberculosis*), fungal (*Aspergillosis* and *Zygomycosis*), and parasitic (*Protozoa*, *Nematoda*, *Cestoda*, *Trematoda* and *Arthropoda*) mites and lice, ticks, miscellaneous arthropod infections in the ostrich have been reviewed by Cooper (2005).

The infection of ratites with Clostridium species is a significant health and welfare issue in ratites. *C. perfringens* is a normal inhabitant of the ostrich intestinal tract (Huchzermeyer 1998), and its identification using faecal culture does not necessarily constitute a diagnosis of clostridial enteritis. However, clostridial enteritis and enterocolitis (Frazier et al. 1993) is a common condition for all ages of ratites (Stewart 1994). Changes in diet and interactions with other digestive system infections and parasites often are contributors to clostridial enteritis in ratites (Shane and Tully 1996). Abrupt dietary changes, starvation, stress, relocation, anthelminthics and vaccinations have all been associated with outbreaks. Ingestion of soil and substrate contaminated with clostridial spores may also be a contributing factor. Acute mortality, occasionally with ante-mortem anorexia and depression, is the hallmark of this infection. Autopsy reveals varying degrees of enteritis associated with clostridial spores, and *Clostridium* spp. can be cultured from the affected intestinal sites of affected birds. Treatment with penicillin and zinc bacitracin has been described (Shane and Tully 1996; Huchzermeyer 1998). Huchzermeyer (1998) reported a vaccination protocol using *C. perfringens* Type B and D vaccines at either 1 week or 1 month, or at 3 and 6 weeks. *C. perfringens* and *Clostridium difficile* are common causes for neonatal ostrich chick enteritis (Shivaprasad 2003). Poor management including poor hygiene, overcrowding, low or high temperatures and excessive handling is the main factor causing bacterial enteritis (Foggin 1992). *C. difficile* may result in enterocolitis and high mortality in young ostriches (Frazier et al. 1993). Establishment and maintenance of normal intestinal flora is crucial.

Adenoviral enteritis has been reported in ostrich chicks in the USA (Raines 1993). Affected chicks show marked depression, anorexia and diarrhoea with mortality usually greater than 90%. Vertical and horizontal transmission can occur. Likewise, Infectious Bursal Disease has been reported in the USA. Affected chicks show depression, anorexia and diarrhoea over a 3- to 4-day period before death (Shane and Tully 1996).

Endoparasitism can be a major health and welfare problem in young ratites, particularly, if reared with adults or if they have access to adult faecal material. An excellent review of ratite parasites can be found in Craig and Diamond (1996). *L. douglassi* (Wire Worm) is a major problem on some ostrich farms on most continents, with mortality rates of affected chicks approaching 50%. It causes a condition referred to in South Africa as "vrotmaag" – rotten stomach with clinical signs including weight loss, depression, anorexia and death (Barton and Seward 1993). Levamisole, fenbendazole and ivermectin have been used to treat infections of endoparasites. Many other helminths have been identified in ostriches and other ratites, including trichostrongyles, ascarids, acanthocephalans and cestodes, but not all of these have been associated with clinical disease.

Prolapses of the cloaca and rectum in young chicks can occur secondarily to enteritis among other causes. Treatment of these prolapses involves replacement with a gloved, lubricated finger and correction of the initiating cause. If the prolapse recurs, the use of a purse-string suture may be required. This suture can be placed under local anaesthesia, and should be left in place for no more than 3 days (Black 1998a).

Treatment of enteritis in ostrich chicks revolves around several key principles:

- Early recognition that a problem exists
- Identification and isolation of affected chicks
- Identification of the causative pathogen
- Identification of environmental, nutritional or management factors leading to the problem
- Appropriate early treatment

Batching of chicks and good biosecurity measures are good initiatives to help prevent and/or limit the occurrence of an enteritis outbreak.

In older birds, cloacal prolapse is commonly found in male and female juvenile ostriches aged up to 6 months (Bezuidenhout et al. 1993), but may also occur in emus (Speer 1996). Clinical signs include severe diarrhoea, intestinal impaction, malnutrition (Speer 1996) and tenesmus (Lumeij 1994; Hoefer 1997). Cloacal prolapse may be associated with *H. meleagridis*, which may occur due to contact with backyard poultry (Iordanidis et al. 2003). Some severe gastrointestinal infections could cause emu chicks to prolapse. It can be treated with Metamucil or mineral oil and antibiotics and the prolapsed portion must be swabbed clean with Betadine and pushed back in through the vent (Minnaar and Minnaar 1995). Samson (1997) recommended a therapy for ostriches that included application of an anti-inflammatory antibiotic ointment, replacement of the swollen tissue into the cloaca under anaesthesia, and placement of a purse-string suture as described for chicks (Black 1998a). While cloacal prolapse is a significant health issue in ratites, it can lead to significant welfare issues for the affected birds (straining, trauma, inflammation, pain and possibly necrosis of the exposed prolapsed tissue).

Mycobacterial infections have been reported in ostriches, emus, cassowaries, rheas and kiwis (Shane and Tully 1996; Doneley et al. 1999; Huchzermeyer 1998; Jakob-Hoff et al. 2000) and such infections have a significant impact on bird welfare. Affected birds typically show chronic weight loss accompanied by other symptoms such as a marked leucocytosis and generally multiple granulomas throughout the liver parenchyma and intestinal serosa, extending into the mucosa. Treatment is often prolonged, difficult and often unsuccessful and there is a limited zoonotic potential; hence, most affected birds are usually euthanased.

Impaction is a significant welfare issue and commonly occurs in ostriches (see Chap. 10) less than 6 months of age (Stewart 1994). It has also been reported in rheas, but is less common in other ratites (Speer 1996) due to the different anatomical and physiological characteristics of the digestive tract, particularly the reduced proventricular curvature and faster gut transit time in species such as emus and rheas (Minnaar and Minnaar 1995). Samson (1997) classified the impactions into acute (weak birds in a few days) or chronic (weak birds in weeks to months), hard (caused by hard materials such as rocks and sand) or soft (caused by fibrous materials such as grass) and partial or complete. Common impaction materials include grass, stones, wet sand, hay (such as long stemmed lucerne or alfalfa) or woody stems, straw and leaves, and even plastic and metallic objects (Komnenou et al. 2003). Impaction is a disease secondary to many other contributory management-related problems such as, when the birds are under stress from high stocking densities, loud irregular noises and excessive human handling (Hicks 1992). Unthrifty and malnourished birds will have a high risk of impactions compared with birds housed under good management (Speer 1996). Good management could reduce impactions by reducing stress factors such as transport of birds around the farm, removing foreign material from the bird enclosure and feeding birds *ad libitum* with a balanced ration. Samson (1997) reported that pantothenic acid deficiency can cause chronic impaction in chicks through excessive ingestion of grass and dirt.

Sick or stressed ostriches of all ages may start to ingest foreign or indigestible material that blocks ("impacts") the proventriculus. As more material (high fibre grasses, stones, dirt, sticks, etc.) is added, the proventriculus becomes distended but no ingesta can move through. Affected birds become anorectic, dehydrated, weak and lose weight. They may regurgitate water after drinking. Faecal output diminishes or ceases altogether. For emus, long fibrous materials such as long grass, synthetic fibres consumed from carpet and string can result in impaction. However, using sand and dirt as floor materials will not necessarily cause impaction in emus (Minnaar and Minnaar 1995). Thus, it is important to identify and correct the underlying contributory management problems that is leading to the impactions. Chang Reissig and Robles (2001) reported gizzard impaction in 1–4 week old lesser rhea chicks in Argentina. Impaction was caused by ingestion of fibrous material, sand, rocks and rubbish (Mushi et al. 1998). Excess fibrous material was also found in the small intestine. Traumatic proventriculitis and air sacculitis in a rhea were caused by the swallowing of metal (Gupta and Trapp 1971).

If one bird suffers from impaction, others in the same pen should be checked. The impacted proventriculus is usually palpable. Undetected ingestion of hard objects could result in gastric stasis. Some objects can be detected by metal detectors and radiographs.

The treatment used in cases of impaction will be determined by the condition of the bird and the nature of the material causing the impaction. For example, birds suspected of eating foreign materials can be fed liquid paraffin or psyllium (Honnas et al. 1991; Frasca and Khan 1997). Tubing with water and fibre laxative (e.g. Metamucil) or mineral oil has been used to treat impaction in emus (Minnaar and Minnaar 1995). Although some authors (Huchzermeyer 1998) advocate against the use of mineral oil, it has been used successfully in sand and gravel impaction when combined with psyllium (Doneley, personal observation). Young birds can be treated by gastric lavage while suspended upside down (Huchzermeyer 1998). If economically viable, some birds may be treated by surgery via proventriculotomy. A proventriculotomy or an esophagostomy can be used to remove the materials for severe impactions (Gamble and Honnas 1993; Samson 1997). Treatment is only of value if the original stressor has been removed, behavioural stress minimised and access to substrate and/or foreign material prevented (Huchzermeyer 1999).

Clinical signs include anorexia, depression and lack of coordination, even death. Ingestion of metal objects such as metallic debris in addition to sand, glass and aluminium can cause sickness or death of birds. Common materials include nails, screws, nuts, staples, pieces of welding rod, battery fragments, coins and lead sinkers. Farmers must site their facilities carefully and attempt to minimise access of birds to metal objects (Glatz 2001). Grazing areas with high levels of heavy metals could also cause heavy metal poisoning in ratites, and for welfare reasons, farming of ratites in these areas should be avoided. For example, high levels of heavy metals and trace elements have been found in ostriches in some regions of the USA.

9.4.3 Diseases of the Respiratory System

The artificial environment for ratite chicks of hatchers, brooders and chick pens sometimes results in poor bird welfare from poor ventilation and the development of respiratory disease. Overcrowding and poor ventilation usually results in a strong smell of ammonia. To improve welfare of birds and prevent respiratory diseases, overcrowding should be avoided (allow 0.5–1.5 m^2 floor space per chick), chick sheds should be cleaned daily to reduce the build-up of faeces and urine and ventilation should be improved. As ammonia is heavier than air (and will therefore settle closer to the floor), ventilation must be provided at chick level.

Aspergillosis is a disease in young ratites commonly resulting from poorly ventilated hatchers or chick rearing sheds. *Aspergillus fumigatus* and *Aspergillus flavus* are commonly isolated from affected birds while *Aspergillus niger* and *Aspergillus terreus* may not be clinically significant (Black 1993a, 1994). When eggs contaminated with *Aspergillus* spp. (from nests) are placed in a hatcher, there is a build-up of spores, which are inhaled by newly hatched chicks. Similarly, chicks placed in poorly ventilated, unhygienic brooder pens may be exposed to spores. If a chick inhales sufficient spores, “brooder pneumonia” may result, often followed by death indicating that aspergillosis is an important health and welfare issue in ratites. In other cases, air saculitis may result, which may not become apparent for weeks, months or even years.

In adult ratites, the most common respiratory disease is air saculitis associated with infection by *Aspergillus* spp. Unlike the "brooder pneumonia" seen in young chicks, this is a chronic disease with clinical signs often emerging within a few weeks of a stressful event such as transport. Affected birds lose weight and show an increased respiration rate (greater than 20 breaths per minute in a non-stressed bird), increased inspiratory effort, abnormal ausultatory signs (inspiratory squeak or grunt and harsh rales) over the thoracic air sacs and open-mouth breathing. Aspergillosis is not contagious among birds, but several birds may become infected from a common source. Treatment with antibiotic (for secondary infections), itraconazole orally and fumigation with enilconazole smoke bombs (Clinifarm, Janssen) may be effective in causing remission of clinical signs in many birds, although recurrence months or years later is common (Love and Gill 1995; Doneley, personal observations). Treatment of severe chronic air sacculitis cases is usually unrewarding because of the granulomatous nature of the disease. Ketoconazole, amphotericin B and itraconazole have been used to treat systemic aspergillosis, while inhalation of smoke generated by enilconazole bombs (Clinifarm, Janssen) has been used to treat pneumonia and air saculitis (Black 1993a, 1994).

Pseudomonas aeruginosa (from contaminated water source), *Pasteurella haemolytica*, *P. multocida*, *Haemophilus*, *Bordetella* and *Mycoplasma* spp. have also been found to cause respiratory infections in ostrich chicks (Huchzermeyer 1998) and in emus and, therefore, need to be considered as health issues that have an impact on bird welfare based on respiratory symptoms.

Avian Influenza, caused by a number of strains of the influenza virus, has been reported in ostriches in South Africa and Denmark and rheas and emus in the USA. Chicks are apparently more susceptible than adults, but all age groups can be affected. Clinically, infected birds show severe depression, respiratory signs, ocular discharge and green urates. Treatment is usually unsuccessful and vaccination relies on the identification of the viral strain present. Prevention is best achieved by preventing contact between ostriches and wild birds (Huchzermeyer 1999).

9.4.4 Diseases of the Yolk Sac

The newly hatched chick receives nutrients and possibly immunoglobulins from the yolk sac, which is attached to the intestine via the vitellointestinal duct (see Chap. 4). Yolk sac contents are drained by both this duct and via absorption by mesenteric blood vessels around the yolk sac. A few days before hatching the yolk sac is drawn into the abdomen and the abdominal muscles close over it at the navel. The yolk is resorbed over 10–14 days, and only a vestigial remnant remains after 3 weeks (Black 1998a).

Occasionally, generally due to incubation problems, a portion of the yolk sac is exposed. If the exposure is only small, dressing it with iodine and bandaging until closure is complete may be all that is required. Larger exposures may require surgical correction if economically practical. Affected chicks need to be monitored carefully for omphalitis or yolk sac retention.

Omphalitis can be a multifactorial problem seen in chicks less than 1 week of age. Chicks that have been stressed by incorrect incubator/hatcher temperature and humidity or chicks that are weak or undersized at hatch or chicks that hatch with an "open" umbilicus or partially exteriorized yolk sac are predisposed to bacterial infection. This infection usually arises from poor hygiene in the hatcher or brooder, although transovarian and transoviductal infections and shell contamination can also be involved. Many chicks that develop yolk sac retention may subsequently develop omphalitis. The incidence on many farms can be reduced by correcting incubator and hatcher problems and improving hygiene. Some farmers routinely apply Betadine to the umbilicus of newly hatched chicks to help to prevent yolk sac infection via umbilicus contamination. If chicks from a particular hen show a high incidence of omphalitis (without other chicks being affected), then that hen needs to be investigated for the possibility of salpingitis/metritis.

On the other hand, yolk sac retention is a failure of the yolk sac to be resorbed in the absence of a primary infectious cause. This is primarily a management problem, with possible faults lying in incubator or hatcher problems, chick nutrition and exercise. Unless secondary omphalitis develops, clinical signs do not become evident until the chick is 10–20 days old (and rarely up to 6 weeks of age) and the autolysing yolk begins to release possibly toxic substances that are absorbed by the mesenteric blood vessels. These chicks fail to thrive and usually start to lose weight. The yolk sac is often palpable in the abdomen, or is detectable via ultrasound.

As antibiotics do not penetrate the yolk sac, treatment for both omphalitis and yolk sac retention is the surgical excision of the yolk sac. Although a relatively simple procedure, the success rate for omphalectomy is not high due to pre-existing toxaemia and immunosuppression.

According to Black (1998a), prevention revolves around:

- Optimal incubation conditions
- Avoidance of early or inappropriate intervention during hatching (an exposed yolk sac has a much higher chance of becoming infected)
- Good hatcher hygiene
- Minimal handling after hatching
- Adequate umbilical care
- Adequate nutrition and exercise

High chick mortality is a significant welfare issue in the ratite farming industries and yolk sac infection is a contributor to the problem. Appropriate management can lead to an alleviation of chick mortality.

9.4.5 Diseases of the Musculoskeletal System

Leg problems in ratites can be readily observed by farmers because affected birds are lame or have difficulty in movement or exhibit an unusual gait – obvious indicators of a health and welfare problem (see Chaps. 4, 5, 8 and 11). Leg deformities, including twisted or bowed leg bones and swollen or deformed hock joints (Speer 1996), commonly occur in young or growing ratites. Tibiotarsal rotation (or angular limb deformity) is one of the most common and serious limb deformities seen in ratite chicks. The tibiotarsus rotates along its long axis, leading to an outward rotation of the limb. Mild cases may show only a "windmilling" gait when running, that is, the affected leg swings out rather than going straight ahead. Severe cases may be so badly rotated that the chick has great difficulty in walking and standing. Numerous causes have been suggested, indicating the problem is multifactorial in nature. Possible aetiologies may include diets excessively high in protein and/or energy, calcium–phosphorous imbalances, leg injuries, lack of exercise and heating of chick pen floors. The greatest likely causative factor in relation to tibiotarsal rotation and limb deformities is inappropriate nutrition. For example, feeding a high-protein starter diet to birds with a lack of exercise results in rapid growth and the potential for leg problems (Jensen et al. 1992; Samson 1997). Although no hereditary component has been conclusively identified, it appears that certain breeding pairs may produce offspring that have high growth rates and so greater potential for growth deformities. These offspring, if fed a diet that can maintain or encourage this high growth rate, will develop a higher incidence of tibiotarsal rotation if they have inadequate exercise. Housing chicks on cement floors with high stocking density contributes to leg problems (Samson 1996, 1997; Minnaar and Minnaar 1995).

Nutrient deficiency could cause leg deformities. Bezuidenhout et al. (1994) found that bone mineralization was poor in deformed bones. Rolled or twisted toes could be caused by deficiency of B-complex vitamins (such as riboflavin deficiency) or unsuitable surface (Huchzermeyer 1994b; Dunn 1995; Deeming et al. 1996) or inappropriate incubation such as excessive humidity. Slipped tendons could be caused by deficiency of manganese (Black 1995; Dick and Deeming 1996). Bow legs are often caused by an imbalance in the Ca:P ratio. Low serum calcium (Chang et al. 1988), selenium deficiency and genetic abnormalities (Stewart 1994) are other factors causing leg problems. "Splayed leg" in emu chicks reported by Minnaar and Minnaar (1995) is related to nutritional deficiencies and injury. In ostriches, "splayed leg" is often seen in weak, oedematous chicks and can be exacerbated by exposure to slippery flooring in the hatcher and brooding area. Deformation of the chest wall with a skewing to one side was found coincident with leg rotations in ostriches (Foggin 1992; Samson 1997).

Stewart (1994) reported that surgery has poor results as a treatment of leg deformities. Various surgical corrections have been described, but none have shown consistent long-term success. Attempts can be made to hobble or splint affected legs but severely affected birds rarely recover and are best euthanased. Rolled toes could be improved by supplying corrective shoes (Wade 1992) or regular nail trimming (Samson 1997). Leg problems can be prevented by feeding young ratites correctly and using good management practices. Samson (1996) concluded that management should be checked and practices changed if over 5% of chicks on a farm suffer from leg problems.

"Wry neck" or torticollis is seen occasionally in all ratite species, with emus being the most frequently affected. The muscles and tendons of the neck are affected, resulting in abnormal positioning of the neck and head. Suggested aetiologies include vitamin E and selenium deficiencies, teratogens, parental malnutrition, excessive handling and turning of the egg or inappropriate temperature and humidity during incubation. Splinting and dietary supplementation may improve many chicks (Speer 1996).

Leg rotation and bruising in ostrich can also occur as a result of poor handling and other trauma. A bird that has suffered leg rotation must be attended to immediately. If the bird has difficulty in rising or walking and has significant heat, pain and swelling, the bird must be euthanased.

Rickets is the result of a deficiency of vitamin D3, calcium or an inappropriate Ca:P ratio . This can be due to dietary imbalances, lack of sunlight, or intestinal malabsorption. It results in enlargement of the joints and bowing of the femur, tibiotarsus or tarsometatarsus (Speer 1996). Early cases may be corrected by identifying and rectifying the inciting causes.

Rolled toes are common for ostriches, but are not as common in rheas and emus (Stewart 1994). This is normally caused by poor management during hatching and early development of young ratites (Speer 1996; Samson 1996, 1997). Rolled toes appear to occur in two distinct groups. The first group is chicks aged less than a week. Within a day or two days of hatching, the toe begins to roll medially. Parental malnutrition and incubator errors have been implicated as causative factors. If treated early, corrective splinting or taping of the affected toe usually rectifies the roll within a few days. Chicks older than 4–5 weeks make up the second group. In these chicks, the rolled toe appears to be a variation of the angular limb deformity problem. Splinting may help with these chicks, but the prognosis is not as good unless dietary and exercise factors are corrected.

"Slipped Tendon" refers to the lateral luxation of the flexor tendon of the gastrocnemius muscle out of its position on the caudal aspect of the hock joint. This condition can be secondary to underlying tibiotarsal rotation. This can be mild, with immediate relocation of the tendon when the joint is straightened, or so severe that it results in a closed or open full dislocation of the hock. Acute, mild cases may be successfully treated with bandaging and splinting. More severe cases, or cases where the luxation has been present for several hours, carry a poor prognosis. Surgical repair of the tendon retinaculum is required in these cases, but is usually unsuccessful because of the weight bearing stresses placed upon it.

Leg abnormalities are a significant welfare issue for ratites. Farmers are often very quick to notice birds that are not walking normally and usually make immediate changes to diet and management in an attempt to correct the problem.

9.4.6 Diseases of the Nervous System

Newcastle Disease, caused by paramyxovirus serotype 1 (APMV1), has been reported in ostriches in Africa and Israel, and in rheas in Brazil. The virus is documented in 236 avian species, confirming widespread susceptibility of birds to this disease (Shane and Tully 1996). Young birds appear to be more susceptible, with up to 50% mortality. Affected birds initially show a slight head tilt, frequent scratching of the head and a tic of the neck muscles. This progresses to torticollis, uncontrolled head movements and finally an inability to lift the head off the ground. Death occurs in 2–3 days (Huchzermeyer 1999). There is no treatment, although some birds recover spontaneously. Vaccination is carried out on some farms in South Africa, with apparently good results (Blignaut et al. 1998; Verwoerd 1998; Schaetz et al. 1998), see the section "Vaccination".

Western Equine Encephalitis (WEE) has been reported in emus and rheas in North America (Vorster and Olivier 1998). Affected birds are initially depressed, progressing to paresis, recumbency and paralysis. Mortality rates are less than 20%, with recumbent birds dying within 48 hours. Vector control and vaccination are usually effective in preventing this disease (Shane and Tully 1996).

Borna Disease has been seen in 2- to 8-week-old ostrich chicks in Israel. Affected chicks show paresis and reluctance to move, progressing to paralysis and finally death from dehydration (Shane and Tully 1996).

Cerebral nematodiasis has been reported in ostriches and emus in North America. *Chandlerella quiscali* and *Baylisascaris* spp. have been isolated from birds showing signs that include ataxia, abnormal gait, muscle weakness, recumbency and death (Craig and Diamond 1996).

9.4.7 Diseases of the Reproductive System

Yolk-related peritonitis occurs when yolk is deposited into the abdomen rather than passing via an intact ova down the oviduct. For example, the infundibulum in ostriches is a large delicate membrane that is quite motile and surrounds a developing ovum on the ovary during the ovulation process. Occasionally, this process is unsuccessful and the ovum fails to enter the oviduct. The exact cause of the failure is often unknown but it may be related to excessive egg-laying, obesity or inflammation of the oviduct or infundibulum. Alternatively, the ovum may successfully become engulfed by the infundibulum and enters the oviduct, but retropulsion pushes it cranially and back out through the infundibulum. The yolk, once released from the ovum after rupture, provokes a peritoneal inflammatory response producing large volumes of fluid as a result. Affected hens usually have a history of having been good layers that suddenly stop laying eggs, or may produced abnormally shaped eggs. They are generally still mating with the male and may not show any other clinical signs. Examination shows distension of the abdomen and a flaccid enlarged vent. A fluid wave can be balloted on cloacal examination, and ultrasound can detect the fluid accumulation. Haematology reflects a marked leucocytosis (Black 2001).

Abdominocentesis, performed with an appropriate catheter (18-gauge spinal needle/catheter in an ostrich) on the apteric region of the lateral flank, reveals a yellowish-pink fluid. Early or mild cases may respond to antibiotic and anti-inflammatory therapy, but severe cases may require a surgical flank laparotomy involving extensive lavage of the coelomic cavity with large volumes of sterile fluids under pressure to remove the yolk material and contaminated abdominal fluid.

Metritis can be due to infection (ascending or descending) or bruising following excessive or difficult egg-laying. The bird may show any or all of the following signs: cessation of egg production, cloacal discharge, stress lines on eggs, increased embryonic mortality, egg retention or signs of ill health. Physical examination and laboratory workup may be unrewarding other than a leucocytosis. Ultrasonography may reveal caseous material and fluid within the oviduct, and an oviduct swab may detect the presence of infection and/or inflammation. Antibiotic therapy should be based on culture and sensitivity, and cases with caseous exudate in the oviduct may require the surgical insertion of a Foley catheter into the magnum or isthmus followed by normograde flushing to expel debris from the oviduct (Hicks-Alldredge 1996).

Egg binding is uncommon in ratites, or perhaps uncommonly diagnosed. With the exception of the kiwi, the size of ratite eggs relative to body size is quite small, and so, the clinical signs associated with egg binding in other species are less pronounced in ratites. The only clinical sign may be cessation of egg-laying. In smaller ratites, the egg may be palpable in the abdomen, but in larger ratites (or cases where the diagnosis is uncertain), ultrasound is usually diagnostic. Conservative treatment with prostaglandin gel (prostin) and calcium may resolve some cases, but surgery may be required.

Although uncommon, prolapse of a large portion of the oviduct of mature egg-producing hens can occur. The prognosis for these hens is generally very poor. Treatment must be instigated as soon as possible after the prolapse and requires general anaesthesia and careful replacement of the oviduct followed by a purse-string suture. If the oviduct has been prolapsed for a significant time, tissue viability is often compromised. Antibiotics and anti-inflammatory or analgesic drugs are given post-operatively. To maintain the welfare of birds and reduce the pain associated with a prolapse, an ultrasound examination is strongly recommended to detect the possibility of the presence of egg retention or other oviduct pathology (Black 1998b).

In contrast to oviductal prolapses, an apparent cloacal prolapse is seen in some immature ostrich hens aged 12–20 months, before egg production begins. The immature oviduct secretes an albumen-like fluid as it matures. This normally drains freely into the cloaca and is occasionally noticed by observant farmers. In some birds, a persistent membrane covers the opening of the oviduct into the urodeum. This prevents the fluid from draining. As pressure builds up behind this membrane, a balloon-like structure arising from the urodeum forms and eventually bulges out through the vent, appearing as a prolapse, 5–10 cm in diameter. Once the diagnosis is confirmed by cloacal examination, the membrane can be incised with a scalpel and the newly created opening widened manually. This allows the fluid in the oviduct to drain and the problem resolves. Prophylactic antibiotic coverage is at the clinician's discretion (Black 1993b, 1998b).

Phallic prolapse can occur in both immature and mature birds. Immature birds, attempting to mate inappropriately, can damage the phallus leading to a prolapse.

As well as the treatment described later, these birds need to be isolated from other birds to allow them time to sexually mature. Mature birds can develop a phallic prolapse following trauma or infection of the phallus and its surrounding structures, or as a result of excessive sexual activity. Occasionally, an unhealthy bird will develop a prolapse due to generalised weakness.

After assessment of the bird via a thorough physical examination, the phallus needs to be replaced and in the case of ostriches, the tip of the phallus sited within the sulcus in the dorsal wall of the vent. Purse-string suture techniques should be a treatment of last resort in these birds. If the phallus is not traumatised or becoming desiccated, conservative treatment is usually sufficient. Isolation and sexual rest, combined with lubrication and frequent replacement of the phallus into the cloaca, may be sufficient. Anti-inflammatory and antibiotic therapy may be useful adjuncts. If the phallus is suffering from significant trauma or drying out, a temporary purse-string suture combined with anti-inflammatory and antibiotic therapy may be required.

9.4.8 Miscellaneous Diseases

In the early 1990s, a syndrome in ostrich chicks less than 6 months old, characterised by weight loss, anaemia and death, started to appear on ostrich farms throughout the world. In Australia, this condition was referred to as “Ostrich Fading Syndrome”. Mortality rates varied from 20 to 80%. Affected chicks progressively lose weight with significant muscle wasting, ascites often develops and urate pasting around the vent is common as the chick weakens. A non-regenerative anaemia is a frequent clinical finding. Autopsy can reveal a multitude of pathologies, including rhinitis, pharyngitis, proventriculitis, enteritis, pneumonia, air saculitis, hepatic abscessation or focal necrosis and splenic atrophy. The most consistent histological finding is a non-suppurative enteritis (Button et al. 1996). The syndrome appeared to have an infectious aetiology and numerous potential pathogens have been recovered from affected chicks including *Cryptosporidia*, gram-negative and gram-positive bacteria, fungi and yeast and a range of viruses. No consistent pathogen has been identified (Button et al. 1996; Speer 1996). Given the wide range of findings and secondary pathogens isolated, it is generally presumed that a primary immunosuppressive agent is responsible for this syndrome. It was suggested that a retrovirus may be the causative agent, but this has yet to be confirmed (Peach 1997).

Erysipelas is occasionally diagnosed in emus in Australia and the USA and is characterised by sudden death mainly during winter. *Erysipelas rhusiopathiae* is transmitted by ingestion of contaminated soil and through skin lacerations. Typically, affected birds die with few clinical signs other than a short period of lethargy and greenish diarrhoea. Necropsy findings included marked petechial haemorrhages on the serosal surface of the large intestine, in particular, pericardial effusion and congestion and mottling of the liver. Culture of *E. rhusiopathiae* from the liver, spleen or heart blood is used as a diagnostic test. Swan and Lindsey (2008) used a treatment consisting of individual or mass medication with procaine penicillin and recommended reduction of stress factors such as overcrowding, and spelling and rotation of paddocks. Isolates from two field outbreaks were identified as strain 21 (Swan and Lindsey 2008). Complete protection was provided by a commercial strain 2b vaccine against challenge by strain 21 field isolates in mice. Annual vaccination of birds at 4 weeks and again at 8 weeks of age appeared to control further outbreaks on farms where the disease had previously occurred and vaccination appeared to protect birds for at least 12 months. Likewise, the turkey vaccine can be used in susceptible birds at 6, 20 and 40 weeks (Shane and Tully 1996).

Salmonellosis has been reported in ostriches, emus, kiwis and rheas. *S. pullorum*, *S. typhimurium* and *S. arizonae* have been isolated, with both vertical and horizontal transmission possible. Carrier birds, rodents, free-living birds and mammals can act as reservoirs for the infection. Clinical signs can include embryonic and neonatal mortality and depression, diarrhoea and sudden death in juvenile and adult birds. Antibiotic therapy (e.g. quinolones) can suppress clinical signs but may create chronic carriers. Good biosecurity measures are necessary to both prevent infection and limit its spread (Shane and Tully 1996).

Anthrax, caused by *Bacillus anthracis*, has been diagnosed in ostriches in South Africa. Spores lying dormant in the soil for many years act as the reservoir for infection. Death occurs acutely with few clinical signs. Demonstration of the characteristic organism in blood smears is diagnostic. There is no recommended treatment. Carcasses should be disposed of by incineration or deep burial. An inactivated vaccine is available (Shane and Tully 1996).

A necrotizing typhlitis, associated with a mixed spirochaete- and trichomonad-like protozoan, has been reported in rheas. Juveniles older than 1 month are susceptible, with affected birds showing depression and anorexia. The mortality rate may exceed 50%. Autopsy shows distension and hyperaemia of the ceca and colon, with fibronecrotic and pseudomembranous changes. Oral metronidazole and parenteral lincomycin reduces mortality in affected flocks (Buckles et al. 1997).

Plasmodium spp. have been reported in ostriches in Africa, rheas in Brazil, and emus in North America. *Leukocytozoon* spp. are considered common in juvenile ostriches in South Africa (Huchzermeyer 1998). An as yet unidentified *Babesia* sp. have been isolated from North Island Brown Kiwis in New Zealand and is considered to be common in the wild population. A *Hepatozoon* sp. have also been recovered from the same birds (Jakob-Hoff et al. 2000). Affected birds of all species show weakness and weight loss, with a regenerative anaemia. Treatment of avian malaria is discussed elsewhere in this text. *Aegyptianella pullorum*, a rickettsia, is transmitted by Argasid ticks in Africa and causes anaemia, pyrexia and death in ostrich chicks. Diagnosis is by detection of the organism in erythrocytes. Treatment with tetracyclines is usually curative (Huchzermeyer 1998).

9.5 Conclusions

The challenge in the ratite industries is to define the extent of the welfare issues that impact on birds as a result of their health. One approach is to develop an indicator of the welfare risk (high, medium or low risk) for health issues. Table 9.1 provides

Table 9.1 Welfare risks (high, medium and low) of various diseases and conditions suffered by ratites based on long periods of stress, pain, debilitation and suffering

Welfare risk		
High	Medium	Low
Clostridial enteritis	Mycotic dermotoses (or low)	
	Skin lacerations (or low)	*Deletrocepahalus dimidiatus*
	Candidiasis	*Struthiolipeurus struthionis*
Libyostrongylus douglassi	Aspergillosis (can be high)	*Gabucinia* spp.
	Omphalitis (can be high)	*Houttuynia struthionis*
Impaction	Yolk sac retention(can be high)	
	Impaction (can be high)	
Aspergillosis	Tibiotarsal rotation (can be high, medium or low depending on severity)	Twisted toes (can be medium)
Omphalitis		
Yolk sac retention	Rickets (medium or low depending on severity)	
	Metritis	Phallic prolapse (can be medium or even high)
	Erysipelas	
	Salmonellosis	
	Anthrax	
	Avian influenza	
	Newcastle disease	
	Clostridial enterotoxaemia	

a guideline for the welfare risks of various diseases and conditions in ratites. Diseases that can result in high mortality are a high risk although high mortality may not be the worst welfare risk as the disease may involve sudden death (e.g. clostridial enterotoxaemia), whereas more chronic diseases (e.g. Aspergillosis and tibiotarsal rotation) could involve long periods of stress, pain, debilitation and suffering and a more significant welfare issue. Table 9.1 provides a guideline for the welfare risks of various diseases and conditions in ratites on the basis of birds being subjected to long periods of stress and pain. Every attempt should be made by the ratite farming industry to avoid these diseases through appropriate management, biosecurity and vaccination. Other health problems and conditions that result in low levels of stress or no mortality may develop into significant health issues if no action is taken to overcome the problem. It is suggested that to improve ratite welfare, welfare audits be developed taking into account major health issues identified in this review.

References

Alexander DJ, Morris HT, Pollitt WJ, Sharpe CE, Eckford RL, Sainsbury RMQ, Mansley LM, Gough RE, Parsons G (1998) Newcastle disease outbreaks in domestic fowl and turkeys in Great Britain during 1997. Vet Rec 143:209–212

Allwright D (1996) Viruses encountered in intensively reared ostriches in South Africa. In: Deeming DC (ed) Proceedings of the 1st international ratite congress: Improving our understanding of ratites in a farming environment, Manchester, pp 27–33

Animal Welfare Advisory Committee (1998) Code of recommendations and minimum standards for the welfare of ostrich and emu. http://www.biosecurity.govt.nz/animal-welfare/codes/ostriches-emus/index.htm#1. Cited 18 June 2010

Anon (1998) Prescribing for birds. In: Bishop Y (ed) Veterinary formulary, 4th edn. Pharmaceutical, London, pp 40–42

Barnett JL, Glatz PC, Almond A, Hemsworth PH, Cransberg PH, Parkinson GB, Jongman EC (2001) A welfare audit for the chicken meat industry. Report to the Rural Industries Research and Development Corporation, Canberra, Australia

Barton NJ, Seward DA (1993) Detection of *Libyostrongylus douglassi* in ostriches in Australia. Aust Vet J 70:31–32

Bezuidenhout AJ, Penrith M-L, Burger WP (1993) Prolapse of the phallus and cloaca in the ostrich (Struthio camelus). J S Afr Vet Assoc 64:156–158

Bezuidenhout AJ, Burger WP, Reyers F, Soley JT (1994) Serum and bone mineral status of ostriches with tibiotarsal rotation. Ond J Vet Res 61:203–206

Black DG (1993a) Aspergillosis in ostriches in Australia. In: Proceedings – Advances in ratite management and medicine, continuing education seminar, A&M University, Texas

Black DG (1993b) Dealing with prolapses in ostriches. In: Proceedings – Advances in ratite management and Medicine. A&M University, Texas

Black DG (1994) Aspergillosis in ostriches in Australia. Australian Ostrich Association Journal 39–44

Black DG (1995) Ostrich examination. Post-Graduate Committee in Veterinary Science, University of Sydney, Sydney, Australia

Black DG (1998a) Chick rearing problems. In: Proceedings of New Zealand Veterinary Association Ostrich Conference, Auckland and Christchurch, pp 20–28

Black DG (1998b) Reproductive disorders of the ostrich. In: Proceedings of New Zealand Veterinary Association Ostrich Conference, Auckland and Christchurch 9:23

Black DG (1998c) Future veterinary roles in the Australian ostrich industry. In: Proceedings Association of Avian Veterinarians – Australian Committee Annual Conference, Canberra, pp 25–30

Black DG (1998d) Nutrition of ostriches. In: Proceedings New Zealand Veterinary Association Ostrich Conference, pp 9–14

Black DG (2001) Ostrich flock health. In: Fudge A, Doneley RJT (eds) Preventative medicine. Seminars in avian and exotic pet medicine. WB Saunders, Philadelphia

Blignaut A, Burger WP, Morley AJ, Bellstedt DU (1998) Antibody production in ostriches in response to vaccination with LaSota Strain Newcastle Disease Virus Vaccines. In: Proceedings, 2nd International Ratite Congress, Oudtshoorn, South Africa, pp 199–200

Bourke M, Glatz, PC, Barnett JL,Critchley KL (2003) Vaccination training manual-edition 1. Publication no. 03/18. Published by the Australian Egg Corporation Limited

Buckles EL, Eaton KA, Swayne DE (1997) Cases of spirochete-associated necrotizing typhilitis in captive common rheas (Rhea Americana). Avian Dis 41:144–148

Button C, Barton NJ, Veale PI, Overend DJ (1993) A survey of *Libyostrongylus douglassi* on ostrich farms in eastern Australia. Aust Vet J 70:76

Button C, Kabay M, Rawlin G (1996) Ostrich fading syndrome in Australia. In: Deeming DC (ed) Proceedings of the 1st international ratite congress: Improving our understanding of ratites in a farming environment. Ratite Conference. Manchester, UK, pp 35–38

Chang Reissig E, Robles CA (2001) Case report-gizzard impaction in Lesser Rhea chicks (Pterocnemia pennata) raised on farms in Patagonia, Argentina. Avian Dis 45:240–244

Chang PH, Chang CF, Liu MRS, Wang KP (1988) Bow leg syndrome in ostrich (Struthio camelus). J Chin Soc Vet Sci 14:17–21

Communities E (1998) Council Directive 98/58/EC of 20 July 1998 concerning the protection of animals kept for farming purposes. Off J L 221:0023–0027

Cooper RG (2005) Bacterial, fungal and parasitic infections in the ostrich (Struthio camelus var. domesticus). Anim Sci J 76:97–106

Craig TM, Diamond PL (1996) Parasites of ratites. In: Tully TN, Shane SM (eds) Ratite management, medicine and surgery. Krieger Publishing, Florida, pp 115–126
Deeming DC, Bubier NE, Paxton CGM, Lambert MS, Magole IL, Sibly RM (1996) A review of recent work on the behaviour of young ostrich chicks with respect to feeding. In: Deeming DC (ed) Improving our understanding of ratites in a farming environment. Manchester, UK, pp 20–21
Dick ACK, Deeming DC (1996) Veterinary problems encountered on ostrich farms in Great Britain. In: Deeming DC (ed) Proceedings of the 1st international ratite congress: Improving our understanding of ratites in a farming environment, ratite conference' Oxfordshire, UK, pp 40–41
Doneley RJT (1996b) Flock health schemes and preventative medicine – the veterinarian's view. In: Ostrich Odyssey, Proceedings Annual Conference Australian Ostrich Association
Doneley RJ (2006) Management of captive ratites. In: Harrison GJL, Lightfoot TL (eds) Clinical avian medicine. Spix Publishing Inc, Florida, pp 957–990
Doneley RJT, Gibson J, Thorne D, Cousins D (1999) Mycobacterial infection in an ostrich. Aust Vet J 77(6):368–370
Dunn S (1995) Ostrich chick rearing. In: Drenowatz C (ed) The ratite encyclopedia. Ratite Records, San Antonio, pp 139–147
Foggin CM (1992) Veterinary problems in ostriches. In: Hallam MG (ed) The TOPAZ introduction to practical ostrich farming. The Producers Association of Zimbabwe, Harare, pp 61–96
Frasca S Jr, Khan MI (1997) Multiple intussusceptions in a juvenile rhea (Rhea americana) with proventricular impaction. Avian Dis 41:475–480
Frazier KS, Herron AJ, Ii MEH, Gaskin JM, Altman NH (1993) Diagnosis of enteritis and enterotoxemia due to Clostridium difficile in captive ostriches (Struthio camelus). J Vet Diagn Invest 5:623–625
Gamble KC, Honnas CM (1993) Surgical correction of impaction of the proventriculus in ostriches. Compend Contin Educ Pract Vet 15:235–244
Glatz PC (2001) Improving skin quality of emus and ostriches. A benchmark study of husbandry, transport, lairage and slaughter methods. Rural Industries Research and Development Corporation Publication No 01/04, ISBN 0 642 58227 0, ISSN 1440–6845
Glatz PC (2008) Ratite toe-trimming-for ostriches and emus-training manual. Rural Industries Research and Development Corporation Publication No 08/017, ISBN 1 74151 606 4, ISSN 1440–6845
Gupta BN, Trapp AL (1971) Case history-traumatic proventriculitis in a rhea (Rhea americana). Avian Dis 15(2):408–412
Hicks K (1992) Ostrich pediatric disorders. Ostrich Report August, pp 13–18
Hicks-Alldredge KD (1996) Reproduction. In: Tully TN, Shane SM (eds) Ratite management, medicine and surgery. Krieger Publishing, Florida, pp 47–57
Hoefer HL (1997) Diseases of the gastrointestinal tract. In: Altman RB, Clubb SL, Dorrestein GL, Queensbury K (eds) Avian medicine and surgery, 1st edn. WB Saunders, Philadelphia, pp 419–453
Honnas CM, Jensen J, Cornick JL, Hicks K, Kuesis BS (1991) Proven triculotomy to relieve foreign body impaction in ostriches. J Am Vet Med Assoc 199:461–465
Huchzermeyer FW (1994a) Ostrich diseases. Agricultural Research Council, Onderstepoort, South Africa
Huchzermeyer FW (1994b) Metazoan parasites. In: Huchzermeyer FW (ed) Ostrich diseases. Onderstepoort Veterinary Institute, Onderstepoort, South Africa, pp 8–45
Huchzermeyer FW (1994c) Ostrich diseases. Agriculture Research Council, Onderstepoort Veterinary Insitute, Onderstepoort, South Africa, pp 11–12
Huchzermeyer FW (1998) Diseases of ostriches and other ratites. Agricultural Research Council, Onderstepoort, South Africa
Huchzermeyer FW (1999) Veterinary problems. In: Deeming DC (ed) The ostrich: biology, production and health. CABI Publishing, Oxon, UK, pp 293–320

Iordanidis PI, Papazahariadou MG, Georgiades GK, Papaioannou NG, Frydas SE (2003) Cloacal prolapse in ostrich chicks with histomoniosis. Vet Rec 153:434–435

Jakob-Hoff R, Twentyman C, Buchan B (2000) Clinical features associated with a haemoparasite of the North Island Kiwi. In: Proceedings Joint Conference Wildlife Disease Association and Wildlife Society, New Zealand Veterinary Association, pp 23–24

Jensen JM, Johnson JH, Weiner ST (1992) Husbandry and medical management of ostriches, emus and rheas. Wildlife and Exotics Animal Teleconsultants, College Station, Texas, pp 124–127

Komnenou AT, Georgiades GK, Savvas I, Dessiris A (2003) Surgical treatment of gastric impaction in farmed ostriches. J Vet Med A 50(47):4–477

Love SCJ, Gill HS (1995) Aspergillosis in ostriches: epidemiology, treatment and diagnosis. In: Ostrich Odyssey 1995, Proceedings 5th Australian Ostrich Association Conference, pp 53–59

Lumeij J (1994) Avian clinical enzymology. In: Seminars. Avian and Exotic Pet Medicine 3, 1–24

Madeiros C (1997) Vaccination of ostriches against Newcastle disease. Vet Rec 140:188

McDougald LR (1997) Other protozoan diseases of the intestinal tract. In: Calnek BW, Barnes HJ, Beard CW, McDougald LR, Saif YM (eds) Diseases of poultry, 10th edn. Iowa State University Press, Ames, USA, pp 890–900

McMillan EG, Zellen G (1991) Histomoniasis in a rhea. Can Vet J 32:224

Minnaar P, Minnaar M (1995) The emu farmer's handbook. NYONI Publishing Co, Groveton, TX

Mukaratirwa S, Cindzi ZM, Maononga DB (2004) Prevalence of *Libyostrongylus douglassi* in commercially reared ostriches in the highveld region of Zimbabwe. J Helminthol 78:333–336

Mushi EZ, Isa JFW, Chabo RG, Binta MG, Modisa L, Kamau JM (1998) Impaction of the stomachs in farmed ostriches (Struthio camelus) in Botswana. Avian Dis 42(3):597–599

Peach P (1997) Detection of retroviral infections and their application to the detection of a novel retrovirus isolated in ostriches diagnosed with Ostrich Fading Syndrome. In: Proceedings Annual Conference Association Avian Veterinary Australian Committee, pp 143–149

Raidal SR, Gill JH, Cross GM (1996) Pox in ostrich chicks. Aust Vet J 73:32–33

Raines AM (1993) Adenovirus infection in the ostrich (Struthio camelus). In: Proceedings Annual Conference Association Avian Veterinary, pp 304–312

Samson J (1996) Targets of performance in ostriches. In: Proceedings of Annual Conference Association Avian Veterinary 14, pp 1–147

Samson J (1997) Prevalent diseases of ostrich chicks farmed in Canada. Can Vet J 38:425–428

Schaetz L, Verwoerd DJ, Grund C, Kosters DJ (1998) Investigation into the Immune Response of Ostrich Chicks after Vaccination against Newcastle Disease. In: Proceedings, 2nd International Ratite Congress, Oudtshoorn, South Africa, pp 195–199

Shane SM, Tully TN (1996) Infectious diseases. In: Tully TN, Shane SM (eds) Ratite management, medicine and surgery. Krieger Publishing, Florida, pp 127–146

Shivaprasad HL (2003) Hepatitis associated with *Clostridium difficile* in an ostrich chick. Avian Pathol 32:57–62

Sotiraki ST, Georgiades G, Antoniadou-Sotiriadou K, Himonas CA (2001) Gastrointestinal parasites in ostriches (Struthio camelus). Vet Rec 148:84–86

Speer BL (1996) Developmental problems in young ratites. In: Tully TN, Shane SM (eds) Ratite management, medicine and surgery. Krieger Publishing, Florida, pp 147–154

Standing Committee on Agriculture and Resource Management (2003) Model code of practice for the welfare of animals. Farming of ostriches. CSIRO Publications, East Melbourne, Australia

Stewart J (1994) Ratites. In: Ritchie BW, Harrison GJ, Harrison LR (eds) Avian Medicine: principles and application. Wingers Publishing, Lake Worth, Florida

Swan RA, Lindsey MJ (2008) Treatment and control by vaccination of erysipelas in farmed emus (Dromaius novo-hollandiae). Aust Vet J 76:325–327

Taylor MA, Hunt KR, Smith G, Otter A (2000) *Deletrocephalus dimidiatus* in greater rheas (Rhea americana) in the UK. Vet Rec 146:19–20

Verwoerd DJ (1998) Newcastle Disease in Ostriches: A review of experimental data. In: Proceedings, 2nd International Ratite Congress, Oudtshoorn, South Africa, pp 192–194

Verwoerd DJ, Deeming DC, Angel CR, Perelman B (1999) Rearing environments around the world. In: Deeming DC (ed) The ostrich: biology, production and health. CABI Publishing, CAB International, Wallingford, Oxon, United Kingdom, pp 163–206

Vorster JH, Olivier AJ (1998) Diseases affecting the central nervous system of ostriches in southern Africa. In: Proceedings of the 2nd International Ratite Congress, 21–25 September, Oudtshoorn, South Africa, pp 201–204

Wade J (1992) Ratite pediatric medicine and surgery neonatal problems. In: Proceedings of Annual Conference Association Avian Veterinary, pp 343–353

Chapter 10
Bird Handling, Transportation, Lairage, and Slaughter: Implications for Bird Welfare and Meat Quality

L.C. Hoffman and H. Lambrechts

Abstract Ostriches have specific behavioural patterns that can be influenced by improper/incorrect rearing conditions. An understanding of these behavioural patterns, as well as how ostriches react under different housing/rearing conditions, will assist the commercial farmer/producer to manage his/her birds to ensure their welfare under *in situ* and *ex situ* conditions. Actions such as handling, transportation, and lairage form part of any commercial ostrich enterprise and can impact on the overall welfare of ostriches. This chapter describes how chicks, juveniles, and adult breeding ostriches perceive their environment, and how changes in behavioural patterns relate to changes in the management programme. Attention is given to both commercial farming and rearing systems. The design of rearing, feedlot, lairage, and breeding camp facilities is addressed to emphasise the importance of facility design in ensuring the welfare of ostriches when handled and maintained. Incorrect facility design can have an adverse influence on meat and skin quality, which ultimately determine the commercial value of a slaughter bird. The chapter also highlights a need for research on various aspects of commercial ostrich farming systems that will assist in optimising the welfare of ostriches of all age groups under commercial farming conditions.

Keywords Behaviour · Handling · Lairage · Slaughter · Stress · Transport · Welfare

L.C. Hoffman (✉) and H. Lambrechts
Department of Animal Sciences, Stellenbosch University, Private Bag X1, Matieland 7602, South Africa
e-mail: lch@sun.ac.za; helet@sun.ac.za

P.C. Glatz et al. (eds.), *The Welfare of Farmed Ratites*, Animal Welfare 11, 195
DOI 10.1007/978-3-642-19297-5_10, © Springer-Verlag Berlin Heidelberg 2011

10.1 General Information

10.1.1 Locomotion in the Ostrich

To be able to minimise the stress that ostriches experience during the herding and transport of these birds, an understanding of their musculature and movement is required (see Chap. 8). Ostriches are strictly bipedal and have been recorded to run up to speeds of 12 m/s with a duty factor (the fraction of a duration of a stride, for which each foot is on the ground) of 0.31 for a female and 0.29 for a male and a maximum stride frequency of 2.3 Hz (Alexander et al. 1979). Ostriches have two major digits (III and IV) with four phalanges on each digit. The last phalanx of digit III, the third toe, is pointed and carries a claw. In some birds, the second toe may have a rudimentary claw. The terminal phalanx of digit I is poorly developed and seen as a minute round bone (Liswaniso et al. 1996). Alexander et al. (1979) assumed that as the third toe is so much larger, most of the force generated when running would act through it and they calculated a force of 1,100 N for a 41.5 kg bird running with a duty factor of 0.29. The third toe is also considered the most significant due to its primary role as a lever for balance, exertion of traction forces, and directional impetus during locomotion (Schaller et al. 2005). Although the ostrich has a thick and strong tarsometatarsus and tibiotarsus in each leg, farmed birds have been noted to snap the tarsometatarsus when running fast. Poor nutrition (i.e. compromise of bone and tendon strength) or spraining of joints (e.g. when stepping in holes when ground is uneven) and the force generated during the latter activity may contribute to fractures that can occur under commercial farming conditions.

10.1.2 Maintenance of Posture

Ostriches normally maintain their balance by standing on both feet, and they seldom will balance only on one foot. When they run, the head and neck are kept fairly upright, and it is only the orientation of the head that will change as the bird maintains visual contact with its environment. Ostriches do have the ability to jump over fences if they are low enough, as well as when they are stressed during handling. Jumping over fences is accomplished by first judging the height of the fence, and then bending the legs at the ankles at a slight angle before propelling them with their feet over the fence. Such a jumping movement normally happens very quickly, and knowledge of the telltale signs will enable a handler to prevent a bird from jumping and potentially injuring itself. The only asymmetrical motion seen occasionally is the 'dancing', 'pirouetting', or 'twirling' performed in certain stress situations (Huchzermeyer 1998).

Having only two major toes per foot plays an important role in helping the bird maintain balance during herding and transport. Toe-trimming of birds adversely affects their balance, particularly when running around corners or when transporting. Toe-trimmed birds are more prone to slipping under these situations (see Chap. 6). Toe-trimming also affects the ability of male replacement breeding birds to create the nest and to mate effectively with breeding females.

Ostriches like to spend a certain amount of time sitting during the day, with most of sitting activity occurring during the night. A sedentary posture is normally assumed when birds settle in to sleep at night, and exposure to the elements can influence the posture that is assumed. When birds sit down to sleep, they sit flat on the ground, with their legs folded under the body and the toes pointing forward. The head and neck are normally laid in a straight line on the ground. When wind is blowing, ostriches will put their head under a wing to protect the eyes from dust particles. The amount of time spent sitting differs between the different age groups, with chicks and juvenile birds spending more time sitting during the day and night relative to breeding birds. Chicks and juveniles tend to sit or huddle together in little groups or crèches, especially during cold weather. The amount of time spent sitting by chicks is influenced by the behaviour and amount of activity of their foster parents. Lambrechts and Cloete (1998) observed that during breeding seasons, females spend more time sitting than males, with the males devoting most of their time to territorial behaviour (patrolling of fences).

10.1.3 Reaction to Changes in Their Environment

The observation of ostriches, in combination with a thorough knowledge of the behaviour of the age group in question, will ensure that potential damage to the animal as well as the handler is minimised to a large extent. Ostriches have conservative behaviour patterns (Lambrechts and Cloete 1998; McKeegan and Deeming 1997) and any sudden change in their immediate environment may cause them to stress and injure themselves. Once a routine has been established in terms of feeding times, for example, this routine should be followed to avoid unwanted stress behaviour in the birds. This is especially important when chicks are raised. It is advisable that handlers and personnel who work with the birds on a regular basis wear the same type of clothing (e.g. blue overalls and a cap), for this will allow the birds to associate them with management-related activities such as feeding or egg collection.

Ostriches have excellent eyesight, which allows them to avoid obstacles when moving about, and they rely on their strong body and weight to overcome obstacles when being chased by a predator or running at full speed (i.e. ±60 km/h). Adult males fight against each other by ramming each other with their chests and by forward kicking. The same kick is also used in defence against other species and predators, including humans. It is not uncommon for male ostriches to run down human intruders, and during handling, a forward kick by an ostrich can cause extensive damage when the claw cleaves open clothing and skin. As the ostrich cannot kick backwards, this is the position from which an ostrich is normally handled to ensure the safety of both the animal and the handler.

Ostriches as a rule start running as soon as they are stressed, and if there is enough space, they will normally accelerate quite fast, should the stress-causing factor continue to be present. Ostriches normally judge the stressor and will slow down if they consider the stressor to be harmless. It is important to indicate the presence of or mark fence wires and posts by means of preferably white indicators (e.g. tyres, metal squares, or PVC pipes). This will allow ostriches to identify the borders of their territories when they are running at full speed, which will then allow them to slow down or change their running direction before they run into the fence.

Failure on the part of the producers to identify the borders of paddocks/camps will result in birds being injured when they try to run through fences when stressed. Douglass (1881) described an incident in which birds were moved from a small yard where they had been raised before allowing the birds to become acclimatised to external stimuli such as sounds and strange sights. When the birds were released into their new environment, it resulted in 12 of the 18 birds being killed as a result of either running into fences or due to exhaustion (i.e. ran for a long period and eventually sat down to never get up again). The latter phenomenon described by Douglass (1881) is most probably capture (exertion) myopathy as described by Huchzermeyer (1998).

Ostriches are also very prone to sudden frights that readily develop into the whole flock stampeding. Triggers for these flights have been found to include hot air balloons and micro-light aeroplanes flying overhead, turbo-driven vehicles passing close to the camp, and cyclists passing close to the camp boundary. In fact any abnormal behaviour that could startle the birds will result in a very rapid flight.

10.1.4 Behaviour

Ostriches have specific behavioural requirements that need to be accommodated in any farming system to ensure that they are raised under as stress-free conditions as possible. Not only is an understanding of the behavioural requirements necessary, but also how the birds will react under different housing/rearing environments and or facilities. Failure to accommodate these requirements may result in abnormal behavioural patterns, which may adversely affect production performance under commercial conditions (Lambrechts and Cloete 1998). By observing the behaviour of chicks, handlers or the producer will be able to get an indication of the well-being of the chicks, as well as make the correct management decisions to address and solve potential problem situations.

Ostriches are by nature very inquisitive and love to peck at or eat foreign objects. Foreign objects may include among other things spark plugs, pieces of glass, and wire, and ingestion may cause puncture of the gastrointestinal tract, which may lead to the eventual loss of birds. Paddocks must therefore be as free as possible from the above-mentioned or other objects that may be detrimental to a bird's health and/or well-being.

10.1.4.1 Routine During the First 6–8 Weeks

It is important to establish a routine during the first 6 weeks of a chick's life. This routine must be followed strictly to minimise potential stress-causing changes in the daily routine. Examples of routine activities include the chick house must be opened every morning at the same time, and chick handlers must wear the same type of clothing, that is for example, caps and blue overall jackets. Chicks tend to associate with the appearance of their handlers (also known as imprinting), and any stranger in their environment may cause them unnecessary stress.

A sense of security is very important for the well-being of chicks, and therefore chicks need to be able to have someone or something that they can focus on as their 'parents'. The type of rearing system, that is, extensive, semi-intensive, or intensive, will determine the type of 'parent figure'. In extensive and semi-intensive systems, the parent figure might be old breeding birds that are used as foster parents. In intensive systems, the chick handler is usually adopted as a parent and plays an important role in the sense of security that chicks will experience. The chick handler will in the latter system usually spend most of his/her day with the chicks. Rabbits and dwarf goats have also been tried as foster parents although with limited success (Huchzermeyer 1998).

10.1.4.2 Learning Normal Behaviour

Exposure to older breeding birds ensures that chicks are taught the right feeding behaviour, which is very important to ensure that the gastrointestinal tract becomes functional as early as possible. Many of the problems encountered, especially in intensive systems and which lead to chick losses, are related to failure of the gastrointestinal tract to function properly. In intensive systems, the chick handler can teach chicks how to eat by playing with the crumbs in the feeding trough. This action will interest the chicks, which are very curious and inquisitive by nature.

Examples of abnormal feeding behaviour include eating of stones and sand, which may cause the impaction of the gastrointestinal tract. It is best to prevent the development of abnormal feeding behaviour through the correct management practices, for once the behaviour has established, it is very difficult to teach chicks to not exhibit this type of behaviour.

Unbalanced diets, too high stocking densities, and sudden changes in management practices may also cause or contribute to the manifestation of abnormal behaviour. Feather pecking is another abnormal behaviour that can be caused by too high stocking densities and unbalanced diets. By bringing the stocking densities down, such behaviour will usually disappear. Unbalanced diets (refer Chap. 6) may cause the development of *pica*, which is the behaviour of eating strange or foreign objects due to a deficiency of certain vitamins and minerals (refer Chap. 5). In the latter case, an ostrich lick can be provided (fed according to the guidelines provided by the processor) to compensate for any such deficiencies.

10.1.4.3 Adaptation to a New Environment

It is important that when ostriches are moved to a new environment, this transition should be made as smooth as possible. For example, birds need to be accustomed to their new feeding troughs and feed at least 2 weeks before they are introduced to their new environment. It is especially important in the case where a change in diet takes place. Ideally, the new diet should be gradually mixed with the old diet over a period of 2 weeks, that is, at the end of the 2 weeks the feed should consist entirely of the new diet.

A change in diet normally occurs when birds are introduced to a new management system, e.g. when they are moved from the growing facility to a feedlot facility, or when they are bought on farm and moved to another. This period is normally characterised by the bird being challenged to adapt to a series of changes, that is, new personnel, new environment (i.e. temperature, noises, outlay of camps, stocking density, etc.), and new management (i.e. handling, feeding, egg collection, etc.). Experiencing some level of stress during this adaptation period is normal, and care should be taken to ensure that the transition between systems is as stress-free as possible for the birds. Stocking density is especially an important factor to consider when birds are moved from a growing facility to a feedlot or quarantine facility, or when breeding birds are moved from a flock breeding system to a system where smaller groups (i.e. smaller flocks, quads, trios, and pairs) are being used. Due to the difference in the design of such facilities, the stocking density usually has to be adapted to be optimal for each system. Too high stocking densities can cause the manifestation of abnormal behaviour such as feather pecking and can also lead to competition at the feeding troughs. Feather pecking can, if the overstocking is not addressed, result in the skins being damaged by sunlight and can therefore contribute to lowering the commercial value of the product. Too low stocking densities is seldom a problem, for the size of the feedlot paddocks is usually large enough to accommodate the running activity of birds that are not normally adapted to being together in groups or used to the presence of farm activities such as tractors driving around. During the time the birds spend in the feedlot, they become more accustomed to the presence of humans and/or other activities, especially during routine feeding and inspection for disease symptoms and/or parasites. This will then ensure that they stress less during the activities associated with the loading, transport, lairage, and slaughter processes.

Orientation in their environment is also very important to ensure the welfare of ostriches. Ostriches have excellent eyesight, which enables them to notice and distinguish objects from afar. It is important that fences are properly marked to minimise damage to birds should they run into fences when scared by any lightning, sudden movements, or predators. Birds when moved to new paddocks or farms are usually disorientated in the beginning, and properly marked fences will assist them to be able to recognise fences in cases where they need to establish borders of their new territories.

10.1.4.4 Aggression

Aggression normally forms part of the reproductive behaviour repertoire of adult breeding males, but can also be encountered in groups of slaughter birds, especially when they are 10 months and older. Ostriches on average enter puberty at 10 months of age. Puberty is the stage in an individual's life when the secretion of the reproductive hormones and especially the levels of testosterone start to increase. Testosterone is the main hormone that is responsible for the development of aggressive behaviour in animals and humans.

Diets high in protein can also increase the level of aggressive behaviour exhibited by slaughter birds. This aggression may manifest itself in abnormal behaviour, should excessive high levels of protein be fed. Aggressive encounters may cause damage to the skins of the birds involved in such encounters, thus lowering the financial value of the skin.

10.1.5 Facilities

10.1.5.1 Farming System

Ostrich farming systems vary from extensive to semi-intensive to intensive systems, with each system having advantages and disadvantages for the birds' welfare.

10.1.5.2 Extensive Farming Systems

In this system, the breeding birds incubate the eggs for the entire incubation period of 42 days and are allowed to raise the chicks until slaughter age.

Advantages

- The most economical option for growing chicks, given that management is such that potential situations and environments that may cause damage to the bird is minimised as much as possible.
- The chicks start to eat as soon as possible and also learn correct eating behaviour from their parents.
- The chicks start to get exercise from the first day of age, which ensures that their legs will become strong at an early age, thus aiding the growth of the chicks.

Disadvantages

- Chicks raised in such a system are unaccustomed to handling and therefore are wild when caught to be transported to the abattoir. This presents a danger for both humans and animals, and may lead to damage caused to the skin and meat, which will consequently decrease the income generated from such a bird. In one such incident when extensively reared birds were transported ~45 km in a properly designed truck on a hot summer morning (>30°C), two birds were dead on arrival and 68 of the surviving 90 birds had to be given an emergency slaughter (i.e. slaughtered immediately upon offloading without any lairage).

10.1.5.3 Semi-intensive Rearing Systems

In this type of system, the breeding birds that were identified as foster parents are allowed to produce a few eggs, which are then removed and replaced with eggs that were already subjected to artificial incubation for 35 days. The foster parents then incubate the eggs for the remainder of the incubation period, that is, approximately 7 days, and thus also hatch the eggs. When the eggs have hatched, the foster parents and their chicks are moved to the growing area, which usually includes a paddock/camp with lucerne pasture and a movable chick house. Chicks are then added to this initial group, anything up to 100 chicks per foster pair.

Chicks are raised by the foster parents up to approximately 4–5 months of age, when they are moved to a feedlot system.

Advantages

- The chicks start to eat as soon as possible and also learn the correct eating behaviour from their parents.
- The chicks start to get exercise from day-old age, which ensures that their legs will become strong at an early age, which will aid in the growing of the chicks.

Disadvantages

- This system when compared to the extensive system is more expensive due to having to maintain the breeding birds. However, the advantages of the chicks learning the correct behaviour from their parents outweigh these costs by far.

10.1.5.4 Intensive Rearing Systems

With this type of system, eggs are incubated and hatched artificially and the chicks are raised in chick houses. The use of chick houses allows for the raising of a large number of chicks on a given area. The design of the chick house will be determined by the area where the house will be built (i.e. in terms of climate, proximity to roads, and other operations, etc.), as well as the raising system that is going to be used.

Advantages

- Large numbers of chicks can be raised successfully if the management of the system is optimal.
- Eliminates the need to maintain breeding birds as foster parents, thus saving on feed costs.

Disadvantages

- The most labour-intensive system, for the houses need to be cleaned on a daily basis to ensure a clean growing environment.
- Personnel are also required to be in the vicinity of the chicks for the entire daylight day. Chicks become accustomed (imprinted) to the presence of a 'parent figure', and absence of such a person may lead to the development of stress-related abnormal behavioural patterns, which may affect their feed intake and thus growth in the long term.

10.1.6 Rearing Systems

The location and design of the commercial operation is going to determine the type of rearing system that is going to be used in growing slaughter birds.

10.1.6.1 Rearing on Concrete/Sand Surfaces (i.e. Intensive Rearing Systems; Week 3 to Approximately 20 Weeks of Age)

Such rearing systems usually consist of a shelter facility, used in conjunction with an outside run. This will allow for the protection of the chicks against the elements, as well as the opportunity to get enough exercise during the day.

Design of Facilities

A variety of shelters are in use in the industry. Examples include custom-built chick houses, modified shipping containers, modified pig or poultry houses, and modified corrugated agricultural sheds.

The type of flooring will depend on the type of housing that is used. The flooring must be of a washable kind or must be replaceable to ensure that the build-up of bacteria and viruses does not occur in the chick house. Interlocking rubber mats and galvanised mesh grids are two popular types of floor cover that can be easily cleaned and disinfected.

The walls of the chick house must also by of such a nature that it can be washed and disinfected with approved and registered cleaning and disinfection materials. Consult with your local co-op on the products registered for cleaning and disinfection that are available in your area and that are suited for use in the specific system in use.

Heating

Young chicks should be raised in an environment with an ambient temperature of 30°C, with a gradual drop of 0.5°C per day, until a temperature of 26°C is reached. Heat sources may include infrared lights or ceramic, oil, or electrical heaters. Care must be taken to ensure that 'hot spots' are not formed. This will have the effect that chicks may sit on each other to conserve body heat, which consequently may cause damage to the skins. Thermometers measuring minimum and maximum temperatures should be placed at different heights to monitor any fluctuations that may occur.

Under-floor heating may be a valuable addition to a chick house, especially if chicks are kept on concrete floors at night or when chicks need to stay indoors for long periods due to bad weather. Temperature regulation is not a critical factor in older chicks. Ostrich chicks can regulate their body temperature from as early as 3 days of age, but this does not mean that they can be exposed to the elements without a negative effect on the chicks. It is recommended that chicks from day old and up to approximately 6–8 weeks of age should be moved indoors when temperatures fall below 20°C.

Ventilation

It is important to ensure that a chick house is adequately ventilated, and to prevent the occurrence of draughts (refer Chap. 6). Adequate ventilation will prevent the build-up of ammonia, which is produced especially during night-time or during the day when the chicks need to be kept indoors due to bad weather. A build-up of ammonia can cause chick losses and must be prevented at all costs. The use of extractor fans will ensure the effective removal of the gases from a chick house. Chick houses, that is, floors, walls, and floor covering, must be cleaned on a daily basis and disinfected prior to new chicks being brought into the facility.

Outside Runs

The nature of the outside runs can be either lucerne/alfalfa-cultivated or concrete-based. The size of the outside run will be determined by the group size. An optimal group size for raising chicks is between 30 and 50 chicks. For chicks aged 4–6 months, a size of 3 × 15 m is usually advised, with shading available in summer. When using lucerne as pasture in the outside runs, it is important to rotate chicks to ensure that they do not ingest roots or long stalks, which may eventually result in the impaction or damage of the gastrointestinal tract.

Feed and Water Troughs

Flat-bottomed plastic containers and poultry-type bell drinkers are usually used during the stage that chicks are raised indoors, that is, for the first 6–8 weeks. After this period, the type of feeding trough and water container should be changed to prevent contamination of the food and/or water, as well as to prevent the chicks from tripping over the feed and water containers.

Self-feeders and truck tyre halves are the preferred type of feeding troughs used in the industry. With self-feeders, it is important to ensure that the feed is protected against rain as well as sunlight to ensure the optimal quality of the feed.

Water should be freely available, and the water system should be cleaned and disinfected at least once during a season (i.e. usually at the beginning of the rearing season) with approved and registered cleaning and disinfection products. Water troughs should be protected against the elements. It is important that the water temperature is never too cold or too hot, preferably between 15 and 20°C. Both too hot and too cold water will inhibit the water intake of the chicks, which in turn will contribute to the potential dehydration of the chicks. Water should also be clean, and water troughs need to be cleaned and disinfected on a regular basis, for example, between cycles of chicks.

Placement of Feeding and Water Troughs

Feeding and water troughs should be placed so as to stimulate the optimal use of the paddock/camp and also to prevent the concentration of the birds in only one area of the camps. At least 8–10 feeding and water troughs for every 50 birds should be provided. From a management point of view, the placement of feeding troughs alongside fences ensures the easy feeding of the birds.

It is important that enough feeding troughs should be placed out to prevent any competition between birds for space at the troughs. Inevitably there are variations in the growth rate of chicks, which will result in chicks of varying sizes in a given group.

10.1.6.2 Rearing on Pasture (i.e. Semi-intensive Rearing Systems; 2–20 Weeks of Age)

Sleeping Facilities

The use of mobile chick houses is advisable for this type of growing system. The reason is to prevent the build-up of potential disease-causing organisms in a given area, which may put the chick's immune system under unnecessary stress.

There are different designs and types of material that can be used in the making of mobile chick houses. Such shelters can include proper buildings with adequate heating and ventilation, field shelter built from straw bales and roofed with corrugated iron sheets, or shelters designed entirely from corrugated iron.

Exposure to Elements

It is important to ensure that the chicks are protected against the elements during the evenings, especially during winter. Areas that are characterised by sudden climate changes will not make good growing areas. The combination of rain and wind can also be fatal due to the wind-chill effect and may contribute to large-scale chick losses.

Pasture

Lucerne/alfalfa is the preferred pasture for growing ostriches. Care must be taken during the hotter time of the year to ensure that young chicks do not ingest wilted lucerne leaves. This may lead to impaction and the eventual death of the young bird. Lucerne that is too old should also not be used as pasture, for the woody parts of the plant may cause damage to the gastrointestinal tract of the young chick.

Growing Paddocks/Camps

Chicks are moved to growing camps at an age of 6–8 weeks, at a density of 75–100 birds per hectare. Shelter should be available to protect the birds against the elements. From this point onward internal parasite control becomes important, for the birds will be exposed to parasites on the pasture.

With this system, it is important to use a system of rotational grazing to ensure that the chicks do not ingest stalks and stems of too old lucerne pasture. As mentioned above, this may cause damage to the stomach and intestines of a bird.

10.1.6.3 Rearing with Foster Parents (i.e. Semi-intensive and/or Extensive Rearing Systems, 2–20 Weeks of Age)

Foster parents are usually old breeding birds that are past their prime and that are culled from a breeding flock (refer Chap. 4). A foster pair can raise up to a maximum of 200 chicks, but the number of chicks will depend on the size of the paddock that is going to be used for the growing.

It is important to remember that when chicks are placed with foster parents, the new chicks appear shorter than the chicks already with the breeding birds. When chicks are taller than chicks already placed with the birds, the male perceives them as being from another clutch and will subsequently kill the new chicks by kicking or trampling them.

A male and female can each protect, that is, cover the chicks at night with their wings, between 10 and 15 chicks from day old to 2 weeks of age. This number will decrease as chicks grow and thus become bigger. The remainder of the chicks should be kept in the mobile chick houses overnight. Foster rearing enclosures must have low wire netting at the bottom of the fence to keep potential predators out, as well as to keep the smaller chicks from escaping.

10.1.6.4 Rearing in Feedlots (i.e. Intensive Systems; Finishing to Slaughter)

This age group is referred to as growers. Growers are moved to feedlots when they reach a live weight of approximately 48–60 kg, that is, at 4–5 months of age. An optimal group size is 50–100 birds per 0.5–1.0 ha. In feedlots, a complete balanced feed is offered *ad libitum* (approximately 2 kg feed/bird/day, depending on diet formulation), and fresh clean water is freely available on a daily basis. Birds are maintained in this system up to the day that they are moved to the quarantine facilities prior to being slaughtered.

Feedlot Facilities

Feedlots can be lucerne paddocks/camps, natural veld areas, or paddocks that are denuded of vegetation. Supplemental feeding is usually provided in the case of natural veld areas. The most common type of feedlot is paddocks that are denuded of vegetation.

Fences should be sturdy and clearly marked, e.g. with white-painted objects such as dropper poles of old tyres. Wire mesh should be used to make the feedlot's perimeter fences predator-proof, for entrance of predating animals into feedlots may lead to large losses due to the birds being frightened and running into the fences. It is advisable to use floodlights at night to illuminate the area to enable the birds to identify the intruder as well as the fences when they start running. Protection in the form of windbreaks should also be available to provide protection against inclement weather.

The feedlot should be a biosecure unit and not located close to the other ostrich-related facilities (e.g. feed mills or chick houses) on the farm and/or poultry operations. Feedlots should also preferably not be located in an area that is characterised as being very dusty, for this may contribute to the occurrence of air sac disease in ostriches.

The location of feedlots should be close to the abattoir to minimise transport-related damage to the skin and meat of the bird. Feedlots should also be close to areas where maize and lucerne is cultivated to minimise feeding costs. Almost 80% of input costs are feed related, and location close to such areas will thus decrease the running costs of a growing facility.

10.1.7 Holding/Quarantine Facilities

Birds are usually detained during five scenarios, that is:

- When chicks are moved from the hatchery to the rearing facility (i.e. between different farms)
- When slaughter birds are moved from the rearing to the feedlot facility (i.e. on the same farm or between farms)
- When slaughter birds are moved from the feedlot to the quarantine facility for the 2-week period prior to slaughter (i.e. on the same farm or between farms)
- When birds are held overnight in lairage at the abattoir prior to being slaughtered
- When breeding birds are moved between farms, for example, after being sold or bought

In the above-mentioned cases, birds are moved between farms or localities, with the detainment being equivalent to a quarantine period to ensure that no diseases are transmitted between farms. During this period they are regularly inspected for disease symptoms and/or the presence of parasites, depending on the situation. In the case where chicks, slaughter birds, or breeding birds are moved from one location to another on the same farm, standard biosecurity guidelines should be adhered to.

The current legislature for the export of ostriches and related products (i.e. meat, leather, and feathers) from South Africa requires that feedlots, as well as quarantine facilities, must be surrounded by a 3 m vegetation-free zone on all sides. This will prevent or minimise the access of rodents and ticks that carry potential disease-causing bacteria and viruses such as the Crimean Congo Haemorrhagic Fever virus. Quarantine paddocks should also preferably be cleaned on a monthly basis, that is, remove faeces and any loose soil. Loading facilities should also be close to the quarantine camps to minimise the exposure of birds to potential vegetation or areas where ticks may occur.

10.1.8 Handling Facilities

There are variations in the design of loading facilities, with producers normally designing a loading facility that forms part of their feedlot operation. There are also a lot of variations and theories among producers as to the best way of loading birds, but the general practice is to have a central medium-sized pen where all the birds to be loaded are herded into. Large pens make it more difficult to catch the birds and will allow them more space to build up speed when fleeing and possibly injuring themselves and causing stress. An optimal pen size is approximately 10 × 20 m. The design of the loading pen is hexagonal to avoid birds getting trapped in corners, which also prevent the birds from injuring themselves.

The location of the handling facility should preferably be far away from roads to prevent the birds from being frightened by people and vehicles passing by.

A handling facility usually consists of the following components:

- *Reception kraal*: This is the kraal that receives incoming birds and in which a bird is being caught before being brought into the main handling kraal.
- *Main handling kraal*: This area should preferably be covered with either a roof or shade cloth to protect the birds and personnel against the elements.
- *Crush/raceway in main handling kraal*: The raceway can be a permanent structure or mobile. The raceway should be wide enough to ensure that a full grown bird can move along comfortably. The sides should be covered; conveyor belting has been found to work well and is very practical (Fig. 10.1). Frequently the raceway would have a raised platform on one side for the handler to walk on whilst herding the birds.
- A restraining box (Fig. 10.2) can be built as part of the raceway or as a separate structure. Two chains fitted with a PVC pipe or a rope or belt usually forms part

Fig. 10.1 A well-designed handling and loading facility. Note the walls covered in conveyor belting to minimise damage to the birds

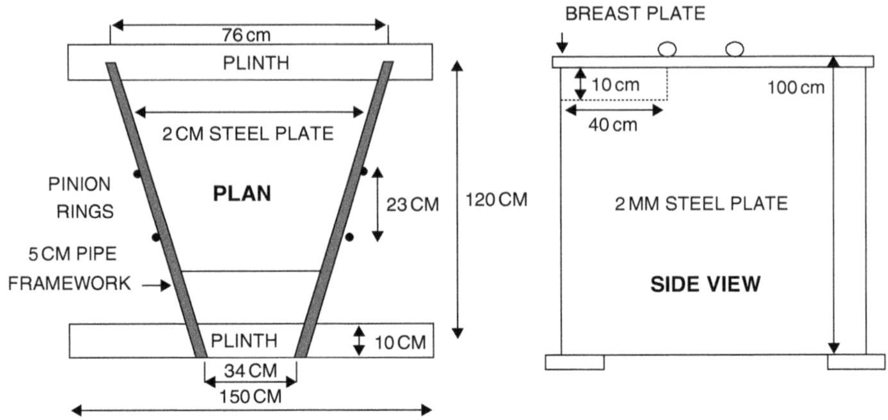

Fig. 10.2 A diagrammatic sketch of a restraining box

of the raceway when it is a permanent structure in the main handling kraal. Both chains are used to keep the bird in place in the raceway, with one chain being placed over the back of the bird and the other under the pelvis bone of the bird. Metal is normally used as material when a mobile raceway is made. Two bottom beams will then form part of such a design to allow for the picking up and carrying of this type of raceway.

- *Weighing area or moveable scale*: A scale consists of two beams (which contain and protect the cables attached to the weighing unit) and a platform that is placed over/attached to the beams. The platform should preferably be a metal plate that is covered by conveyor belt to absorb the sound when a bird steps onto the platform. Platforms that only consist of metal will get hot during summer, especially if the facilities are not shaded during summer.
- *Sorting paddocks*: The number of sorting paddocks will depend on the size of the commercial setup, but usually there is anything from four to ten paddocks.

Refer to Appendix for two examples of handling facility design.

The following should be addressed in the design and manufacture of the handling facility:

- The fences of the receiving paddock and the main handling pen should be sturdy enough to take the pressure of birds running or pushing against the fence. The fences usually consist of tar poles covered with conveyor belting to prevent any damage to the skin of the birds should they run into the fence.
- Fence height should preferably be high enough to prevent the birds from looking out, that is, ±1.6 m. When Kenyan Reds or Zimbabwean Blues are part of the farming operation, the fence height should be increased to accommodate the greater height of these breeds (i.e. ±1.9 m).
- Fence material should not consist of diamond mesh fencing or barbed wire, for ostriches tend to get stuck in this type of fence.

- There should be no corners in these paddocks, for this will increase the chances of the bird and/or handlers being injured during the catching of the bird. Ostriches when cornered will attempt to jump or turn around and kick, which is dangerous for both the animal and the handlers.
- Fence poles should be on the outside of fences, for ostriches like to run alongside camp boundaries. Care should be taken that there are no protruding wires or loose ends of conveyor belting. Protruding wires can injure the bird when it runs into it. Ostriches when being handled experience high levels of stress and usually look for a way out. In cases where there are openings as a result of conveyor belting not attached properly, a bird may stick its head through such an opening and eventually may end up injuring itself.
- Gates that close off the accesses or entrances of the pens and paddocks should be affixed properly and of the generic kind. Gates should not be covered with shade cloth or corrugated iron plates, for birds are frightened by both types of cover on gates.

10.1.8.1 Loading Facilities

The loading facilities should be designed to allow the transport truck access to the area where the birds will exit the holding pens. The level of this passage is to be on the same height as the floor of the truck to allow birds to walk into the truck compartments. Loading facilities were designed in the past to incorporate a ramp, but this is not advisable for birds that do not like to walk uphill, especially when being hooded and they cannot see where they are going.

When using a loading facility where a ramp forms part of the design, the ramp incline should not be too steep, for ostriches do not like walking up an incline. The walkway should also be wide enough to allow for the bird as well as a handler on either side to fit comfortably on the walkway. When a ramp is designed, it should be kept in mind that a bird may struggle when it is led onto the ramp, and therefore, the width of the walkway should accommodate this.

When loading ramps are made out of metal, a covering such as a thin layer of conveyor belting should be used to cover the surface of the ramp. This will absorb the noise that is created when a bird steps onto the ramp, which will help to minimise the noise stress that is experienced during handling. The conveyor belting will also prevent the metal from heating when loading activities are performed during a hot day. The latter scenario is the exception to the rule, for normally loading and offloading should be limited to the cooler hours of the day to ensure the least amount of heat stress to the birds.

Typically, the loading facility would be situated in a corner of the pen and consist of a crowd/forcing pen, a raceway, and then the loading ramp (Fig. 10.1). All the sides should be solid from leg height upwards to ensure ventilation as they keep the birds calmer and facilitate movement. The sides should be high enough (1.7–2.0 m) to prevent the birds from seeing distractions. A walkway on one or both sides will allow a stockman access to the birds. The crowd gate that closes behind the birds should also be solid. There should be no sharp corners or protrusions as these will cause injury and bruises. The floor of the crowding pen and raceway should be coarse and slip-proof to minimise injuries should a bird fall and is normally sand, free of stones, etc. that may cause damage to the feet of the birds. Care should be taken, however, to ensure that the sand that is trodden out is replaced as birds, especially birds that are not used to humans and being handled, have been noted to lie down so as to allow the crowding gate to pass over them.

10.1.8.2 Design of the Transport Truck

When a contractor is used to transport chicks or slaughter birds, ensure that the contractor and his personnel have experience of transporting ostriches. The design of trucks and trailers that are used to transport ostriches must adhere to certain requirements to ensure the safe transport of ostriches (Fig. 10.3).

These requirements are as follows:

- The sides of the truck should be solid (i.e. with no gaps at the bottom) and an opening between the roof and the side to ensure adequate ventilation. Most trucks that are used in the South African industry to transport ostriches are, however, not fitted with roofs, so ventilation is generally not a problem.
- The floor of the vehicle must be solid, non-absorbent, and slip-free. Typically metal gridding or rubber matting is used. This will prevent birds from slipping and the latter will also provide protection against the cold metal surfaces, especially if long distances are going to be covered. Although rubber matting is less durable, observations and discussions with transporters

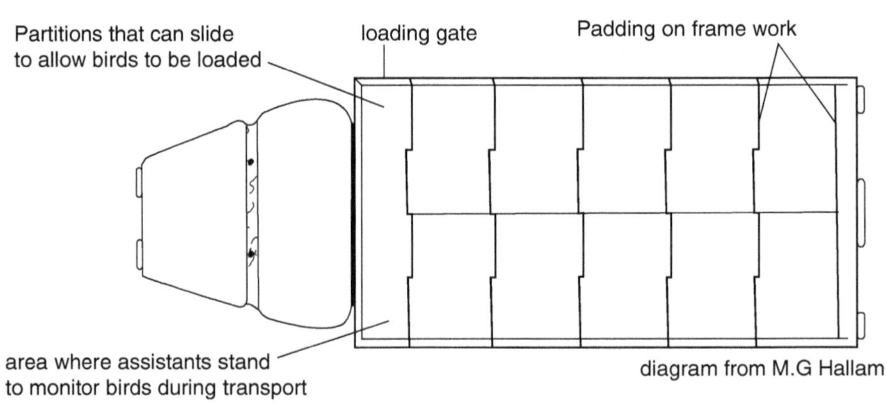

Fig. 10.3 Design of a truck to be used for the transport of ostriches

indicate that this material is more suitable as the metal mesh can hurt the toes of the birds.

- The corners of trucks need to be protected with a form of cushioning material to prevent the birds being injured by any sharp edges in the corners. The loading space may not have any other sharp angles, protrusions, or holes that may injure the birds. No loose objects should be stored in the truck compartments.
- Partitions must be installed at every 3 m of loading space if the truck is longer than 4 m.
- The minimum required floor space per bird is 0.5 m^2.

Using the optimal stocking density when transporting birds is very important to ensure as stress-free a journey as possible. It is also important to remember that stocking densities should be adjusted when transporting ostriches during summer, when birds need more space and ventilation to be able to thermoregulate properly.

It is important to use enough handlers per compartment to monitor the status of the birds during transport, that is, whether they are sitting down or trampling each other. Birds that tend to sit down during transport are frequently hung in a sling harness in one of the corners of the compartment or removed and placed in a special transport box that is usually situated between the front and rear tyres of the vehicle.

When driving, it is important to maintain a speed that will not endanger the birds and handlers when the truck needs to stop quickly. Acceleration and braking should be smooth, and sharp turns or tight corners should be avoided when transporting birds.

10.2 Handling, Restraint, and Transport of Ostriches

10.2.1 Handling of Ostriches

Handling of ostriches should be limited to periods when management-related activities need to be carried out. Such activities may include vaccination, treatment against internal and external parasites, diseases, and weighing. Ostriches should always be treated with the utmost care and respect, for they are quite unpredictable under certain circumstances and can inflict serious damage. Unnecessary handling may cause stress that can impact negatively on the growth and welfare of the animal. Personnel to handle the birds need to have experience in the handling of ostriches. Care needs to be taken to ensure that neither the personnel nor the animals are stressed due to improper handling techniques.

It is important to remember that the handling of ostriches will never be a completely stress-free exercise; care must thus be taken to minimise potentially stressful conditions. Using properly designed handling facilities when handling ostriches of any age group will eliminate some of the stress that accompanies handling (see the earlier section on handling facilities for more details on the requirements of handling facilities). The correct design and use of the facilities will prevent injury to the handlers and birds.

Personnel who are going to be involved in the raising and handling of the chicks and growing birds need to be familiar with the normal behavioural patterns and body language of ostriches of the age group in question. For example, chick behaviour is considered to be a good indicator of a chick's well-being. To accustom chicks to being handled from an early age will make them easier to handle as they grow older and will also potentially minimise stress behaviour during handling. Chicks that are not used to being handled or to the presence of humans become wild and a danger to themselves and their handlers when handled later on during their lifetime.

10.2.1.1 Handling of Chicks from Day Old to Approximately 3 Weeks of Age

The rearing of ostrich chicks can be divided into three stages, that is, immediately post-hatch, from day old to 4–6 months of age, and finishing in feedlots. *Immediately post-hatch* is the most critical stage in the rearing of ostrich chicks. Chicks should be kept indoors for the first week, in an environment of which the temperature is controlled and ventilation is adequate to remove the ammonia that is produced. During winter times, chicks can be kept indoors for up to 2 weeks before being allowed to venture outside. A space requirement of 0.16 m^2 per chick is recommended and should be extended by 10% on a weekly basis. It is extremely important that the chicks are not overheated or become too cold. Overheating will result in the chicks developing diarrhoea. Observing the behaviour of the chicks will indicate whether they are overheating. Chicks will pant and sit with their wings spread open if they are too hot. Too cold ambient temperatures will lead to poor yolk absorption and a higher susceptibility of chicks to contract diseases. Chicks that are too cold will sit on each other, which may eventually cause damage to their skins (Engelbrecht et al. 2009). After this period, chicks are usually placed with foster parents or moved to the growing system with outside runs.

After the first week, chicks are allowed to venture outdoors during daytime, weather permitting. It is advisable that chicks are raised in chick houses for up to 6 weeks (summer) or 8 weeks of age (winter) before they are allowed to sleep outside at night. A group size of 30–50 chicks is recommended, for this allows for easy management and early detection of any health-related problems.

It is important that chick handlers and growers are trained how to handle day 1 and older chicks to ensure that neither the handler nor the chick is injured or stressed during handling. In intensive rearing systems, it is advised that chicks are handled frequently, for example, once per month when being weighed, to allow them to become accustomed to handling and the presence of humans.

To catch chicks younger than 3 weeks, it is important that they are not picked up by the neck or legs, as is the practice with poultry chicks. Chicks of this age group tend to run for a short distance and then crouch down. This is their way of trying to hide themselves from the handler. Chicks can be gently picked up and put into a crate and carried to the area where they are going to be dosed or vaccinated.

10.2.1.2 Handling of Chicks from 3 to Approximately 16 Weeks of Age

Chicks older than 3 weeks and up to approximately 4 months of age need a different handling method. The best method to handle chicks of this age group is to place one hand around the base of the neck, and one hand under the abdomen, and then lift the chick up. The legs are usually left to dangle free.

As the chicks grow in size, handling the birds as stress-free as possible may require two or more people. The one handler will usually catch the bird by the neck, just below the head with the one hand, and then use the other hand to take hold of the lower jaw. Care should be taken to ensure that the pressure on the lower jaw is not too high; otherwise the thin fleshy part may be perforated and therefore cause the bird to be injured. It is good practice to ensure that a second handler is always nearby when a chick is caught. Sometimes a chick, especially as they grow older and heavier, may be difficult to handle. The second handler will then take hold of the pelvis bone of the bird and help to steer the chick in the direction of the handling facility. If there is a shortage of personnel, a chick's head can be covered by a hood that will quiet down the bird. However, a chick from this age group seldom likes a hood being placed over its head, and it is advised that the handler pull the chick's head slightly down so that it cannot focus directly on its environment and thus remove the chick's ability to struggle or kick forward.

10.2.1.3 Handling of Ostriches 4 Months and up to Slaughter

During the finishing stage (i.e. from approximately 4 to 5 months to slaughter), juvenile birds are raised in feedlots or grazing systems. It is especially important to remember with these older chicks that care should be taken to not stand in front of the chick whilst catching or handling the bird. Ostriches kick forward, and a kick from a heavy slaughter bird (i.e. approximately 100 kg) can cause severe injury. It is best to stay at the side or towards the back end of the bird when handling older chicks.

When working with birds of 4 months and older, it is advised to work with them in large groups, that is, as much as the handling facility can comfortably contain without putting too many birds in the handling kraal. The behaviour of the group should be observed, with any highly stressed or aggressive birds being caught first to prevent agitation and thus increasing amounts of stress in the group.

Individual ostriches can be captured by using a shepherd's crook or by catching the beak in one hand and then pulling the ostrich's head down and then pulling in the direction it is to be moved. Care should be taken that the finger/thumb is not

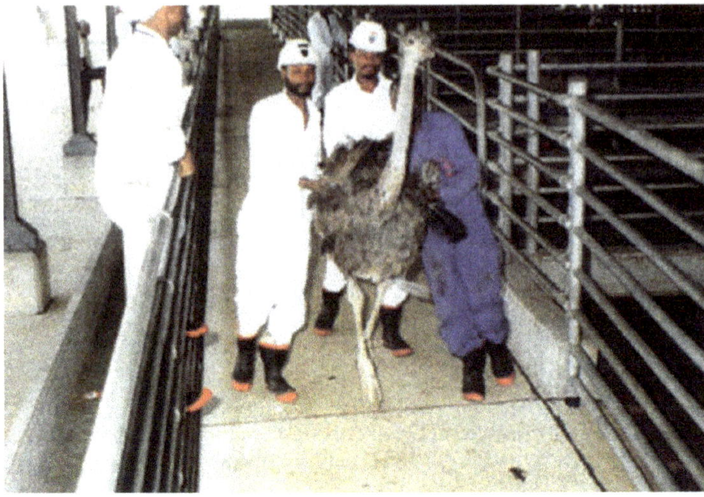

Fig. 10.4 Three handlers moving a bird – one on each wing and the third pushing/raising the tail

forced through the lower mandible (beak). Alternatively, the ostrich (with or without a hood) is forced by one person on each side holding the upper wing close to the body in the auxiliary region and one behind the rump pushing the tail up into the desired direction (up or down the loading ramp, into the stunning box, etc.) (Fig. 10.4).

During the action of catching the bird, care must be taken to avoid its forward kicks. Lifting the tail end up and holding the head down makes it more difficult for the bird to kick. Care must also be taken not to exert excessive force to the wings of the bird and to hold them close to the shoulder joint to avoid their fracture or dislocation (Huchzermeyer 1998).

10.2.1.4 Handling of Breeding Birds

Breeding birds should preferably be handled before and at the end of a breeding season to ensure that their normal reproductive behavioural repertoire and activities are not disrupted. Handling of breeding birds during these sessions normally involve treatment for internal and external parasites, vaccination against Newcastle Disease Virus, and other actions such as collection of semen for evaluation and ultrasound scanning of females to determine the amount of follicular activity on the ovary. Breeding birds are much heavier and taller than juveniles, and personnel must have the proper training and experience to handle breeding birds. This is especially important when Zimbabwean Blue or Kenyan Red ostriches form part of the breeding flock, for both genders of both breeds are on average heavier and larger than South African Black males and females. Kenyan Red ostriches are also more aggressive by nature, especially after they have entered puberty, and personnel must be made aware of this and trained to handle them to limit the amount of stress the birds will be subjected to during handling.

It is of the utmost importance that when breeding birds are translocated to a new breeding environment, this activity is performed during the non-breeding season. Breeding birds will need time to acclimatise to their new environment, which will ensure that reproductive activities and behaviour are optimal in the new environment. It is recommended that breeding birds are made accustomed to their new diet/feed 2 weeks prior to translocation to ensure that their feed intake is not affected. Different management practices should also be introduced to ensure that the breeding birds will not be stressed by any such changes in their new environment. Failure to do so will impact negatively on their reproductive activities and performance.

Breeding birds normally become accustomed to their breeding territory, and should they be moved to a new camp in a successive year, this can cause their reproductive performance to decrease by as much as 50%. The number of eggs produced by a female is highly repeatable (i.e. the number of eggs produced during a breeding season will give a fairly accurate estimation of the number of eggs that will be produced in the following season), and it is therefore important that allocation to a breeding environment should be performed to keep uprooting and translocation to a minimum. The influence of translocation should be kept in mind when females are evaluated for reproductive efficiency. Should moving a female be inevitable, it will take at least two breeding seasons for her egg production potential to be restored.

It is advisable to have a team designated to only catching birds, that is, these handlers will have the necessary experience of how to single out and catch birds with the least amount of stress. They will then pass the bird on to the team that is designated to work with the bird, for example, weigh, vaccinate, or treat the birds for internal and external parasites.

The best procedures for the catching and handling of slaughter birds are as follows:

- Single out a bird from the group, normally the catching team will be the only people in the pen. The bird is usually caught by means of a shepherd's hook (Fig. 10.5), which has a long handle (about 2 m long) and a specially shaped hook that is designed to be wide enough to accommodate the width of the neck of the bird (i.e. slaughter as well as breeding birds). Sometimes if the handlers are experienced enough, they can jump up to catch the bird by the part of the neck just below the head; this should, however, only be done with the necessary experience.
- As soon as the bird is caught in the hook, another handler should position himself or herself behind the tail end of the bird to prevent the bird from running in a circle or reversing, which will make it difficult to remove the head from the hook.
- Immediately after the handler has positioned him/herself behind the bird, the head must be taken out of the hook, and a hood placed over the head to quieten

Fig. 10.5 A bird being led by means of a shepherd's hook. Note that the neck is pulled downwards

the bird down. A hood is normal 15 × 30 cm and should have an opening large enough for the beak, which will also allow the bird to breathe comfortably. The opening should, however, not be too big; otherwise the bird will still be able to observe its surroundings and start running when not being handled properly. Sometimes handlers use old socks as a hood, for example, in the case of younger birds. This is inadvisable as due to the elasticity of the sock material, birds cannot breathe comfortably.

- After the hood has been placed over the bird's head, two handlers position themselves at the wings (i.e. on either side of the body) to then help to steer the bird to the handling crush or scale, depending on the activity that is going to be performed (Fig. 10.4).

Care must be taken that the noise level is kept as low as possible when handling or working with chicks and/or slaughter birds. Ostriches are very sensitive to the atmosphere in the handling facility and will become difficult to handle if noise levels are excessively high.

Handlers should always work quietly and calmly during the herding and loading of the birds. The use of electric prodders or hitting the bird with a stick against the neck should not be allowed. Hitting against the neck can result in a bird losing its balance and falling over backwards, which can result in it injuring its neck to the extent that it has to be culled. This can especially happen when breeding birds and more specifically males are handled at the onset of the breeding season, that is, when they are more aggressive than usual. The use of whips is also not advised, for they tend to get entangled in the bird's legs and cause them to fall, injuring themselves and possibly also the handlers (Douglass 1881).

10.2.2 *Restraint in the Ostrich*

The objective of restraint of an animal is to minimise the danger to the handler and animal, as well as to minimise the stress to the animal. Typically, each animal has a specific manner in which restraint is applied to attain these goals.

The method used to restrain ostriches is influenced largely by the age and physical size of the bird, as well as the reason for restraint. For example, a day-old chick can be managed by a single handler, whereas the handling of adult birds requires three to four people. All personnel working with chicks, juveniles, and/or breeding birds should be trained in handling techniques, as well as behavioural aspects. Handlers should be able to judge the stress level of the birds and adjust their handling methods accordingly. In South Africa, a routine loading and transport (in this case ±65 km) of slaughter ostriches to a local abattoir during the summer resulted in a 15% mortality due to the excessive heat stress experienced by the birds when inexperienced stockmen were used to load and transport the birds. Transporting the birds during high ambient temperatures (>30°C) increases the number of birds that die due to their inability to adequately thermoregulate.

When ostriches are restrained, it is important to use handling facilities that are properly designed to allow for the handling procedure to be carried out with the least amount of stress to the animal. Unsuitable handling facilities and inexperienced personnel can contribute extensively to wounds or bruises on carcasses. In the case of juvenile or slaughter birds, this is especially an important factor to consider, for any bruising or wounds will result in the eventual downgrading of the skins and meat, which will result in a loss of income.

10.2.2.1 Restraining the Bird in a Crush

After a bird has been caught, the bird is directed into a crush and positioned so that its breast bone rests on the breast plate (see design in Fig. 10.2). Different types of materials such as wood, rubber, and metal can be used to construct a crush. Normally a crush is made from wood or metal, with the latter being the type of material used for mobile crushes. When wood is used, a crush normally consists of three 1.5 m poles of approximately 150 mm diameter, which are firmly planted 400 mm deep into the ground. The two sides of the V should be 1.2 m long and the opening 0.9 m wide. The top of the sides of the V is closed in with sturdy planks (1.2 m × 300 mm × 30 mm), which are fastened to the inside of the upright posts with recessed bolts. The inside front and sides of the crush are padded with rubber, leather, sheep skins, or hessian to prevent abrasions (Fig. 10.1). Provision is made for a pole to be pushed through behind the bird's legs to keep it in place and on the sides there are rings for fastening a strap over the bird's back and under its wings (Huchzermeyer 1998). The bird can then be handled manually as described into the crush.

A crush is also used to restrain birds during periods when feathers are harvested and is thus also known as a plucking box. Once the bird is comfortably positioned, the two fastening attachments, that is, one over the back and one under the tail end, are placed in position and the bird is secured in the crush. Depending on the design of the crush, that is, a crossbar (slipped in under the tail end) and a rope (fitted over the back) or chains covered with PVC piping can be used to secure the bird.

Care must be taken not to use the wings to force the bird into the crush. By doing so, one or both of the wings might be broken, which may eventually impact on the bird's display of reproductive behaviour or thermoregulatory ability. Sometimes a bird may reverse if the tail end is released too quickly, which may result in the bird, especially a juvenile slaughter bird, going through the open front end of the crush. Handlers should preferably not release the bird before it is secured.

10.2.3 Loading of Ostriches

Ostriches that are transported by means of a truck normally include juvenile or grower birds (5 months and older), slaughter birds (±10–14 months of age), and breeding birds (2 years and older). Loading can be a dangerous activity if inexperienced personnel and/or inadequate facilities are used.

The stress experienced by the ostriches during the loading phase will be influenced by the level of training of the stockmen handling the birds, as well as the design of the loading facility. Birds that are habituated to humans are more tame and easier to handle and are less prone to stress in any new environment. Wilder birds may be transported with hoods on. It is also advisable to transport birds of the same size together and not to transport male birds in their breeding season as they are very aggressive and tend to injure themselves and other birds.

When herding a group of ostriches into the catching pen, it is important to ensure that the group is neither too small nor too big. Too big a group will result in the handlers not being able to catch a bird without getting injured, and too small a group will give the birds an opportunity to run around too much and thus increase the chances of being injured when they build up speed and run into a fence.

Birds are caught by means of the procedure described above. One bird is caught and loaded at a time, and if two catching teams are used, this will ensure for a more rapid loading tempo. It is also advisable to place a hood over the bird's head to quiet it down as much as possible.

When loading juvenile or slaughter birds, it is advisable to adhere to the compartment dimensions, that is, do not load more birds than the number allowed per compartment. Preferably load birds of similar size together in a compartment. Larger birds in a compartment may be more aggressive and are prone to kick or injure smaller birds in the compartment. The opposite is also applicable, that is, do not load too few birds per compartment, for this may cause them to fall around during the journey and injure themselves. Falling of birds will cause bruising of the carcass and will damage the skins, which will result in poor welfare as well as downgrading of both products, and a loss of income in the case of slaughter birds.

When loading breeding birds, the size and gender of the birds should be kept in mind when allocating birds to a compartment. Males and females should be transported separately. During hot and/or humid days, it is advised to allocate fewer birds to a compartment to ensure that the birds will be able to dissipate body heat effectively. When birds are overweight, it is advisable to allocate fewer birds to a compartment to allow the birds to not overheat during the journey.

Birds do not walk freely up a ramp onto a truck and therefore have to be manhandled onto a truck. When a handling facility that involves a ramp is used, two to three handlers are normally used to load a bird, that is, one on either side and one at the tail end of the bird. The neck of the bird should not be used to steer the bird in the direction of, and into, a compartment. Ostriches do not like their neck being touched, and doing so during a loading activity may result in a bird reversing off the ramp. Care should be taken to ensure that the truck is parked correctly and that there is no space between the loading ramp and truck where a bird's feet can be caught. Frequently the placing of an additional wooden plank across any opening ensures that it is safe to load the birds.

When a handling facility without a ramp is used, that is, where the loading surface is level with the floor of the truck compartment, care must be taken that the birds are not chased into the compartment. Birds should be allowed to walk into the compartment. Pressure can be applied at the back of a group to ensure that the pace, at which compartments are filled, does not take too long, as this may cause stress to the birds that were loaded first – this is particularly applicable on a humid and/or hot day.

The raceway is normally wide (3 m) as very few birds will allow themselves to be herded individually. The bird handler normally stands on the walkway and when a hood is used, the bird would typically be caught from this position and the hood placed over the head. It is common to have a number of hooded birds standing together in the raceway waiting calmly for their turn to be loaded onto the truck.

10.2.4 Transport of Ostriches

10.2.4.1 Land Transport of Ostriches

Stressors arising during transport were broadly categorised by Crowther et al. (2003) as either 'irritant' or 'intermittent'. Irritant stressors are those most likely to occur over prolonged periods, for example, continuous noise, vibration and movement, heat exposure, and confinement in a novel environment; continued exposure to 'irritant' stressors tends to have a compounding effect, which with time can lead to undesired levels of stress. Intermittent stressors occur less frequently and provoke a response from the birds that is immediate and of very short duration. During transit these may be shafts or flashes of light caused by passing cars or street lamps, or noise from other vehicles passing by. The latter was also noted in video recordings by Mitchell (1999) where birds showed stress behavioural responses linked to visual external stimuli; of note was the negative reaction of the birds to the approach of humans. Both types of stressor may have implications on the overall welfare status of ostriches; however, it is the 'irritant' stressors that appear to have the greatest effect on elevating stress levels during transportation. During the transportation of ostriches, the vibrations of the vehicle as well as the environmental temperature were shown to enhance the skin temperature and modify heart rate (Crowther et al. 2003). Mitchell (1999) showed that stress during transportation caused changes in blood chemistry similar to that associated with physiological stress, which may include fatigue, dehydration, and tissue damage. The transportation induced a 1.7-fold increase in plasma glucose, indicating stress-induced mobilisation of glycogen reserves and gluconeogenesis. This was linked to a 50% reduction in plasma lactate (thus indicating the substrate for the later). Kamau et al. (2002) evaluated the effect of mixing and transporting juvenile ostriches by examining the blood heterophil to lymphocyte ratio and found that the ratio increased – a clear indication of stress.

A decrease in glycogen reserves in the bird *ante-mortem* will result in a high muscle ultimate (after 12 h *post-mortem*) pH (pH_u). Muscle with a pH_u of >6.0 is defined as being dark, firm, and dry (DFD). DFD will cause the muscle to have a strong water binding capacity and the muscle will appear unattractive. The high muscle pH will also cause a shortened shelf life due to bacterial growth.

Schaefer et al. (1995) showed that the pre-transport application of electrolyte therapy reduces the losses in both live weight and carcass weight and suggested that pre-slaughter stress can modify meat colour, pH, and drip losses. This was quantified by Fasone and Priolo (2005) who showed that heavily stressed birds (e.g. birds that had broken legs or were stunned incorrectly) had higher ultimate meat pH (6.95 vs. 5.91) and darker muscle (*M. iliotibularis*) that also showed a higher water binding capacity (as indicated by a lower cooking loss).

It was found that during the night transportation of birds, their heart rate and skin temperatures were lower – the latter may be linked to the cooler ambient temperature during the night (Crowther et al. 2003). During periods of darkness birds chose to sit which may be a diurnal/nocturnal response or simply an instinctive reaction to darkness. The sitting response resulted in a lowering of heart rate. The increased stability associated with sitting and the decrease in potential for heat stress strongly suggests that the practice of transporting ostriches at night is advantageous from an animal welfare perspective.

Immediately after transportation an ostrich will excrete a thick white concentrated urine characteristic of dehydration. This will cause fouling of the lairage floor and, depending on the floor type, could result in contamination of the skin and feathers.

10.2.4.2 Sea Transport of Ostriches

Transporting ostriches, and more specifically breeding birds, by sea is not a common practice, but has been used especially when birds are translocated between continents. During sea transport, ostriches are usually housed in containers that are specifically modified to accommodate feeding and water troughs, as well as equipped with special flooring material to prevent the birds from slipping.

As is the case with transport on land, transport by sea is also a stressful experience. During a sea journey where breeding birds were exported from South Africa to Dubai, the birds developed extreme aggressive behaviour, abnormal homosexual behaviour, as well as stargazing behaviour (Pfitzer and Lambrechts 2001). It is advisable that during the sea transport of ostriches, male and female breeding birds should be transported separately to minimise aggressive encounters between males and females. To prevent the manifestation of stargazing behaviour, it is advisable that modification to the design of containers includes routes by which the birds can be exposed to daylight. Sufficient exposure to daylight is needed to prevent the occurrence of this behaviour.

Haloperidol, a neuroleptic used to calm down game animals during capture, translocation to, and acclimatisation in their new environment, was used successfully to quiet or calm down birds during the above-mentioned journey. It is important that the administration of tranquilisers should always be performed by a veterinarian who is familiar with the drug and its side effects, as well as normal behaviour of breeding ostriches.

10.3 Lairage of Ostriches

The objective of a lairage for slaughter animals is to allow them time to settle down after a stressful journey. Initially, it was proposed that a lairage time of a minimum of 24 h would allow most animals sufficient time for their muscle glycogen levels to be replenished. This would then allow normal muscle metabolism (anaerobic) *post-mortem* resulting in good quality meat. However, it is now accepted that 24 h is insufficient time for most red muscled animals to replenish muscle glycogen levels to that found in rested muscles. The time spent in lairage is now shorter and is determined more by management requirements than any other reason. For example, ostriches may only spend a few hours or up to 24 h in lairage depending on the slaughter rate and throughput in the abattoir. It is also accepted that the lairage *per se* is a stressful environment with various inputs such as proximity to other flocks, new flooring, new pens resulting in a higher density of birds, unknown noises, etc. all resulting in additional stress. The effect of these multiple inputs on the meat quality has been little researched in ratites.

Typically when birds arrive in lairage after transport, they would stand around and defecate. As the birds become calm they will start to lie down. In some abattoirs therefore, the practice is to offload birds into external lairage pens with sand floors for 2–4 h. When the birds have finished defecating and have settled down, they are normally moved to a smaller lairage pen having sand, cement, or a grid floor, which is normally under a roof. The birds are kept in this pen until moved off to be slaughtered. These pens are all hexagonal in shape, thus stopping the birds from crowding into corners. Water is always freely available to the birds.

Ostriches defecate more readily during penning (Burger et al. 1995), and the subsequent spoilage of the hides is one of the main contributors of bacteria of faecal origin on ostrich meat. At South African abattoirs, the birds also have unrestricted access to drinking water (Van Schalkwyk et al. 2005); too much water leads to an increase of alimentary tract volume, which complicates evisceration and often leads to contamination of carcasses through rupturing of the full intestines.

Research at a South African export approved abattoir by Burger et al. (1995) evaluated the difference in microbiological quality of ostrich meat after lairage of the birds standing on two different types of surfaces. In this study, two groups of ostriches were kept for approximately 24 h in lairage at the abattoir; one group in pens with clean river sand as flooring; the other in pens with cement flooring. The ostriches were slaughtered under identical conditions and meat was sampled in overnight cold rooms. No statistically significant differences were found between the aerobic plate counts on the meat from the ostriches penned on sand or cement.

A number of abattoirs have a cement floor with grooves built into the floor to stop the birds from slipping. However, this was seen to be unsuitable as the birds tended to slip and a metal grid (1–1.2 cm thick rods, 5–6 cm^2) was then placed above the floor to minimise slipping. However, it was then noted that the birds would not lie down as the metal grid would hurt their feet. A floor system that has been found to be highly efficient is made of a metal mesh (1–1.5 cm sided square holes) raised above the floor. All the excreta then fall through and are removed by daily flushing of the floor underneath.

Van Schalkwyk et al. (2005) reported on the effect of feed withdrawal (feed deprived) during lairage on meat quality characteristics in ostriches. After evisceration, the mass of the full stomachs and the stomach contents of the stressed groups (feed deprived) was found to be lower than that found for the control group, but the mass of the full alimentary tract and the alimentary tract contents were slightly higher for the stressed group (no significant variance in any of the weights). It was thus suspected that feed withdrawal will reduce the risk of carcass contamination at evisceration due to decreased viscera volume that prevents the puncturing of the intestines. There was, however, a significant difference in intramuscular pH between the control and the stressed groups in the study of van Schalkwyk et al. (2005). At 1 h *post-mortem*, the readings of the stressed birds were 0.22 units higher, and after 26.5 h in the cold room, the readings were 0.25 units higher than the control. These high pH values (between 6.03 and 6.46) in the stressed group could make the meat of the stressed birds more susceptible to microbial growth and could be indicative of meat with a shorter shelf life.

Fasone and Priolo (2005) reported that ostriches which had been stressed from both transport and lairage practices had a significantly higher ultimate pH (6.95 vs. 5.94) than the unstressed control group. This corresponds well with results found in practice for birds delivered stressed at the abattoir where the research for the rest of this study was conducted. The unstressed control group values reported by Fasone and Priolo (2005) correspond to those reported by Sales and Mellett (1996) and Paleari et al. (1995, 1998). Crowther et al. (2003) reported that ostriches are markedly less stressed when transported at night rather than during the day. On the basis of the above-mentioned data, it can be assumed that proper management of transport and lairage practices to minimise stress on the birds will result in a lower ultimate pH in the meat and better holding quality. Very little research was found in the literature on the effect of transport practices on ostrich meat quality or microbiology and this field requires attention.

10.3.1 The Causes of Bruising and the Influence on Microbial Load

Ostriches are often transported over long distances to slaughterhouses and the transporting on trucks, the on- and offloading from the trucks, and lairaging at the abattoir have proved to be the most common causes for bruising on livestock carcasses (Grandin 1990, 1991). In addition to the usual hazards for livestock transportation, ostriches have the added disadvantages that they are bipedal, have two-toed feet, and a high centre of gravity, which all contribute to ostriches having trouble in keeping their balance on the trucks (Wotton and Hewitt 1999). Ostriches therefore have a tendency to sit down during transport, which may lead to severe injuries due to trampling in the confined truck compartments or in the pens.

Producers, transporters, and abattoir management in South Africa should adhere to strict animal welfare codes (SAOBC 2001) regarding the treatment of ostriches during transport and pre-slaughter practices to prevent unnecessary bruising or damage to the skin and carcass. The preventative measures during transport include keeping the birds calm, keeping to prescribed numbers of birds per truck partition, having handlers travel with the birds on the trucks, and designating experienced drivers for the trucks. Furthermore, the trucks, loading areas, and pens are constructed with rounded corners, no protruding elements, and slip-free flooring. Despite these measures, Wotton and Hewitt (1999) reported that lacerations and bruises on the necks and lower legs were common on ostriches delivered to South African abattoirs.

Wotton and Sparrey (2002) reporting on these precautionary measures taken during transport and handling at a South African abattoir highlighted the serious damage that can be inflicted to both skins and meat by kicking, bruising, or fresh wounds. They reported that ostriches with fresh wounds would often be returned to the farms to heal.

In Chambers et al. (2004) with reference to all livestock species, including ostriches, on the effects of stress and injury on meat quality, it was indicated that because of glycogen depletion during transport and pre-slaughter stress, there is little lactic acid production in the muscles that caused the meat pH to be higher than ideal. This higher pH would then better support microbial growth and the meat from animals that were stressed, injured, or diseased before slaughter will have a shorter shelf life. The FAO document indicated that bruised meat is wasted due to aesthetic unacceptability to consumers and the fact that it decomposes and spoils rapidly due to the bloody meat that is an ideal growth medium for bacteria. This is then the reason for the removal or trimming of these bruises during primary meat inspection; this practice, if not well controlled, can also lead to unforeseen losses in meat yield.

Ostriches are slaughtered and the carcasses de-feathered, skinned, and eviscerated (Hoffman et al. 2006); thereafter the carcasses are inspected. This process is known as the primary meat inspection and is performed in the export abattoirs by Department of Agriculture meat inspectors, appointed under the authority of the Regulations under Act 40 (Anonymous 2004). While inspecting the carcasses for bruises and injuries, the inspectors must trim away visible bruises according to the appropriate Veterinary Procedural Notice (Anonymous 2007). All these actions take place on day 1, within 1 h *post-mortem*. This action of warm trimming of bruises has in the past been known to contribute to significant losses in meat yield per carcass (on average 300 g per bird) (Hoffman et al. 2010). In Table 10.1, where the distribution of bruises on carcasses is noted, it is clear that a high incidence of bruising occurs on the neck (53%) of the birds, with the front of the thighs being a second prominent area. The bruising on the necks is caused by the birds rubbing/bumping their necks on the top rails of the transport trucks – an indication that the design of the trucks may be inadequate or incorrect. The bruises on the thighs are typical of birds bumping into objects, whilst large and multiple bruising was probably from the trampling of birds lying down. Hoffman et al. (2010) also established that trimming bruises on warm carcasses caused higher total aerobic viable counts on the trimmed surfaces than cold trimming. Cold trimming together with better management of trimming practices also led to a decrease in meat yield losses.

Sabbioni et al. (2003) noted that lairage (2–26 h) time had a significant effect on carcass weight. Lairage time was also found to affect the *M. fibularis longus* fat

Table 10.1 Distribution of bruises on ostrich carcasses slaughtered over an 8-day period in a commercial abattoir (Hoffman et al. 2010)

Number of			Number of bruises on					Percentage (%) of bruises on			
Day	Birds	Bruise	Neck	Back	Thigh front	Thigh back	Total	Neck	Back	Thigh front	Thigh back
1	300	19	4	0	12	3	6.3	1.3	0.0	4.0	1.0
2	594	190	101	2	66	21	32.0	17.0	0.3	11.1	3.5
3	558	237	123	6	83	25	42.5	22.0	1.1	14.9	4.5
4	335	72	45	1	25	1	21.5	13.4	0.3	7.5	0.3
5	200	58	21	0	28	9	29.0	10.5	0.0	14.0	4.5
6	371	51	27	0	21	3	13.7	7.3	0.0	5.7	0.8
7	248	60	40	0	17	3	24.2	16.1	0.0	6.9	1.2
8	547	108	57	0	42	9	19.7	10.4	0.0	7.7	1.6
Total	3,153	789	418	9	294	74	25.21	52.58	1.13	36.98	9.31

content and fat energy/total energy ratio. The authors attribute this increase in fat content to dehydration caused by stress. The pre-slaughter rest also influenced the fat quality (reduced saturated fatty acids and polyunsaturated fatty acids) and increased the meat's sensitivity to oxidative stress – a result most probably linked to the change in fatty acid profile. Van Schalkwyk et al. (2005) monitored the effect of 2.5 days (a period simulating lairage of ostriches arriving at the abattoir over the weekend and being slaughtered on the Monday) of lairage on the meat quality of ostriches and found that the main effect was on the live weight change (the birds that had not had any feed lost 3.2 ± 0.56 compared to 1.0 ± 0.51 kg). The weight of the hot and cold drumsticks was independent of the treatment. Treatment did have an effect on the *M. iliofibularis* pH post-mortem with the birds being without feed for the longest period having the highest readings. There were no differences on the physical quality attributes (drip loss, cooking loss, and shear force) of the meat.

Lairage is stressful for birds, not only because of a new unknown environment but also because of the activities that occur in and around the lairage and abattoir areas. In an experiment conducted in our laboratory, the effect of lairage time on the muscle quality of 78 ostriches from the same flock was evaluated. After spending the night in lairage, 38 of these birds were randomly selected and killed immediately when the slaughtering commenced (early) at 06:30 in the morning. Killing of these birds was completed by 07:15. The remaining 40 birds were kept in the same lairage pen and were the last group (late) to be slaughtered that day. Slaughtering commenced at 16:00 and was completed within 45 min. The design of the lairage facility is such that the birds were kept in a pen adjacent to the stunning pen. This meant that the early group was not subjected to the herding of foreign birds past the pen as were the late group. All animals had unrestricted access to clean drinking water throughout the experimental period. Although no differences were found in the water binding capacity of the muscles between the two groups, the birds slaughtered later had darker coloured muscles – an indication of stress. These birds also had higher initial muscle temperatures, as well as high muscle pH values (Fig. 10.6). Both these phenomena are indicative of *ante-mortem* stress.

10.3.2 The Effect of Slaughtering and Dressing Techniques on the Meat Quality

Restraint and stunning require specialised facilities and procedures because of the long neck, head anatomy, and physiology of ostriches. An electrical current in excess of 400 mA at 50 Hz AC applied only to the head would prevent recovery in more than 90% of the ostriches when bled within 60 s of stunning (SAOBC 2001). It was also noted that the first stages of recovery in the birds are accompanied by rhythmic breathing movements. While observations of breathing could be a diagnostic tool on the effectiveness of the stunning, identification of rhythmic breathing movements in the ostrich after stunning is difficult because the spinal

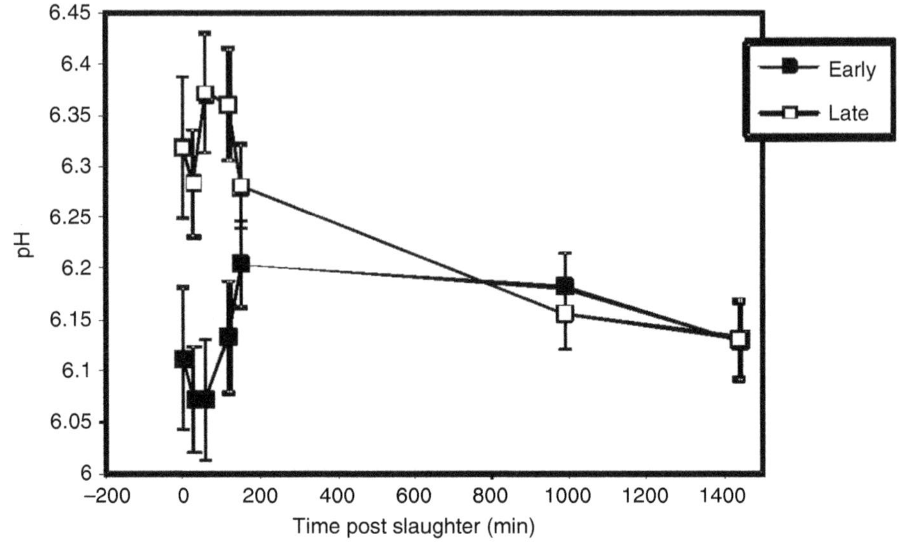

Fig. 10.6 Effect of position on the slaughter line on the post-mortem pH of fillet samples of commercial slaughter ostriches, taken 0, 30, 60, 120, 150, 990, and 1,440 min after slaughter. Vertical bars about the means denote standard errors

reflexes causing limb muscle contraction also result in almost rhythmic body movements that could be confused with breathing movements. The South African legislation requires a current of 400–600 mA, 90–110 V for a duration of 4–6 s.

Stunning has traditionally been with hand-held tongs as birds are held in the restraining area by pressure normally applied by gently pushing from behind on the tail feathers. The area is often a V-shaped structure, high enough that the stunning operator is not kicked. After (and sometimes during) stunning, the birds are rocked backwards and a rubberised leg clamp placed over the legs at the tarsometatarsal bone, thereby immobilising them, and allowing the birds to be ring/chain shackled via the big toes. Birds are hoisted onto a 3.4 m overhead rail and manually conveyed to another area for exsanguination. This conventional stunning procedure has been replaced in many abattoirs with a new restraining and stunning mechanism that completely encompasses the ostrich in a padded clamp holder (Hoffman 2005). Double padded sides restrain the bird's upper thighs and a rubberized foot clamp holds the feet so that there is no physical damage to the bird. As the bird is electrically stunned, the entire stunning box rotates 180° so that toe clamps can be applied without any danger to the stunning operators. The restraint is opened after stunning and the bird is hoisted and conveyed for exsanguination.

Within 20 s of stunning, the birds should be bled by means of a complete ventral cut to the neck and/or by thoracic sticking. The head is normally held between two horizontal metal bars to minimise blood spillage on the feathers and skin. After stunning ostriches are bled by means of a complete ventral cut to the neck and/or by thoracic sticking (TS). Although no research has been reported on the effectiveness of these two methods, personal observation seems to indicate that better and faster bleed-out is obtained when both the neck cut and thoracic stick are performed. In a pilot investigation in which the two bleeding methods were applied, ten ostriches subjected to ventral throat cut alone had an average (±s.d.) bleed-out percentage (defined as weight of blood expressed as a percentage of live weight) of 2.8 ± 1.03%. When the ventral cut was combined with thoracic sticking, the bleed-out was 3.3 ± 0.34%. The means could not be proved to differ significantly ($P = 0.16$), but the magnitude and direction of the absolute difference seem to warrant further investigation (Hoffman et al. 2009).

Lambooij et al. (1999a, b) evaluated the effect of different electrical and mechanical (captive needle pistol using air pressure) stunning procedures on their efficiency and effects on some meat quality parameters. The rigor mortis value in the tenderloin (*M. ambiens*) and the pH_1 (45 min *post-mortem*) and pH_2 (18 h *post-mortem*) in the big drum (*M. gastrocnemius*), tenderloin, and triangular fillet (*M. ilio-femoralis*) muscles were lower ($P < 0.05$) when stunned with air pressure compared with electrical stunning. These authors also noted that a short stun-stick interval (5 s vs. 39 s) results in lower pH_2 values in the tender loin and triangular fillet muscles and a better water binding capacity in the big drum. They recommended that at least 500 mA be applied and to use a short stun-stick time interval or to kill the birds by a long stunning duration. They also noted that the captive needle pistol, using air pressure, can be an alternative for electrical head only stunning (Lambooij et al. 1999a).

Wotton and Sparrey (2002) noted that an electrical current in excess of 400 mA at 50 Hz AC, applied to the head only would prevent recovery in more than 90% of the ostriches when bled within 60 s of stunning. Wotton and Sparrey (2002) also noted that the identification of rhythmic breathing movements indicates the first stages of recovery in the birds and could be a diagnostic tool in recognising the effectiveness of the stun. However, as they also noted, the identification of rhythmic breathing movements in the ostrich after stunning is difficult because spinal reflexes, which cause the contraction of limb muscles and result in almost rhythmic body movements could also easily be confused with breathing movements. The South African legislation requires a current of 400–600 mA, 90–110 V for duration of 4–6 s. Bleeding (by means of a complete ventral cut to the neck and/or by thoracic sticking) should be achieved within 20 s of stunning.

These three papers all conducted research where the birds were stunned with hand-held tongs and the birds were maintained in the stunning 'box' in a gentle manner (pressure normally applied by pushing from behind via the tail feathers). This stunning box consists of a V-shaped metal structure, high enough to ensure that the stunning operator will not be kicked. The birds are then pushed into the closed corner formed by the V-shaped structure. After (and sometimes during) stunning, the birds are rocked backwards and a rubberised leg clamp is placed over the legs at the tarsometatarsal bone, thereby immobilising them, and allowing the birds to be ring/chain shackled via the big toes. The birds are then hoisted onto a 3.4-m overhead rail, manually conveyed to the point of slaughter where a high neck cut is performed. The head is normally laced between two horizontal metal bars to minimise blood spillage on the feathers and skin. This stunning procedure has since changed in most of the major abattoirs in South Africa with the use of a new stunning box.

The box design allows for the whole bird to be restrained. The Divac Ostrich Stunning box (Divac, PO Box 257, Knysna, 6570, Republic of South Africa) is built from a combination of galvanised mild steel and stainless steel, which encapsulates the ostrich in a padded clamp type holder, ensuring no physical damage occurs to the bird. The bird is gently pushed into the box, which is then closed manually around the bird. The double padded sides restrain the bird by holding the upper thighs. A rubberized pneumatic foot clamp restrains the bird's feet whilst the bird's head is placed manually into the stunning clamp. As the bird is stunned (using the standard device as already discussed), the whole stunning box rotates through 180°, thereby positioning the bird for the toe clamps. This restraining device allows the placement of the toe clamps without any danger to the operators. After the stunning is completed, the box is opened and the bird hoisted for further processing (exsanguination, etc.). The time from stunning to exsanguination has also been reduced to <20 s. Once released from the clamp, the unit rotates to its initial position, ready for the next ostrich. The effect of the use of this clamp on the meat quality is still to be quantified. However, the quality of the feathers and the skin is not compromised in any way, as was the case using the old leg clamp method of restraining.

After stunning the birds are bled by means of a complete ventral cut to the neck and/or by thoracic sticking. The effect on ostrich muscle quality of an additional thoracic stick (TS) to the normal ventral throat slit to bleed ostriches after electrical stunning was evaluated (Hoffman et al. 2009). The additional TS had no negative or positive effect on the drip loss, cooking loss, colour or pH, and temperature readings of the fillet (*M. iliofibularis*), big drum (*M. gastrocnemius*, *pars interna*), and inside loin (*M. iliotibialis cranialis*). Nonetheless, personal observations would recommend the use of TS as it seems as if the ostriches die faster.

Some abattoirs in South Africa electrically stimulate (ES; 45 V, 0.4 mA; 10 s on, 10 s off for 3 min) the carcasses during the bleeding phase. It is postulated that this helps with the bleed-out and thereby increasing the shelf life of the meat. However, no research has yet been reported to substantiate this. Morris et al. (1995) noted that ES 45 min post-mortem had no influence on muscle pH and temperature decline. However, we now do know that ostrich muscle pH undergoes a very rapid decline post-mortem (Botha et al. 2004a, b) and if any benefit was to be derived from ES, it would most probably only be of value if it was applied as early post-mortem as possible. An early post-mortem low voltage electrical stimulation (ES) of the carcasses also had no influence on the cooking loss, drip loss, and colour of the fillet (*M. iliofibularis*), big drum (*M. gastrocnemius*, *pars interna*), and inside loin (*M. iliotibialis cranialis*) (Hoffman et al. 2009). Electrical stimulation did result in a lower pH_{45} in both the fillet and big drum muscles. However, after 24 h, the pH of the muscles did not differ. Electrical stimulation also caused elevated initial muscle temperatures, although this effect was only temporary due to the efficient cooling mechanism used in the abattoir. Electrical stimulation also had no effect on the Warner Bratzler shear force values in the fillet. It can be concluded that low voltage

ES has no advantage pertaining to physical quality characteristics of importance in an ostrich abattoir.

Severini et al. (2003) also noted that a pre-de-feathering shower did not affect the carcass microbial load, but a final carcass washing had an effect. Although the authors do caution that their sample numbers were too low to determine a definite effect. Unfortunately, the authors do not indicate what form of water was used or whether any additives had been added. The aspect of final carcass washing warrants further research.

10.4 Conclusion

Although adequate guidelines are available to ensure that ostriches are handled humanely, and that the stressors that they are exposed to are minimised, there are still a number of issues that need to be researched and addressed. The whole production system as pertaining to the keeping of brood flocks, hatching of the eggs, and rearing of the chicks seems to be done on a humane and acceptable manner. However, a major issue within the production chain is the high mortality experienced when chicks and young ostriches are transported. Anecdotal information places these mortalities as high as 65%, although this has never been substantiated.

Similarly the weight loss that slaughter birds experience en route to the abattoir needs to be elucidated. Another aspect that warrants research is to qualify and quantify whether the stress experienced during the transport of birds is mainly of a psychological and/or metabolic nature. However, most trucks used for the transportation of birds to the abattoirs seem well designed and suitable for this species. An aspect that warrants further research is the bruising experienced on the necks from birds trying to see out over the sides/panels of the trucks. Nutritional interventions that could help minimise these stressors need to be researched. Most of the lairage designs are such that the stress experienced is minimal and the new stunning box has ensured that the stress experienced by the birds is minimal. This new box also has the advantage that is visually less disturbing than the older method.

Appendix

Design of Handling Facilities

Design 1

- This type of design allows for the sorting of slaughter according to weight or gender. Sorting kraals (A) is also fitted with their own gates as indicated in the diagram to ensure the effective opening of the kraals.

- Handling facility for ostriches that incorporates a revolving gate (B) to manipulate the size of the catching area (A) as ostriches becomes smaller. The handling area (C) is where the necessary equipment such as a scale and crush is placed, and this is the area where all evaluations/administration of medicines are performed. Processed birds are then allocated to the sorting camps (D), which simplifies the moving of birds to specific camps or for culling purposes. Broken lines indicate gates, and thick lines indicate where conveyor belting should be placed to minimise damage to birds during the handling process. The sorting camp fences are the same as for breeding camps (Fig. 10.7).

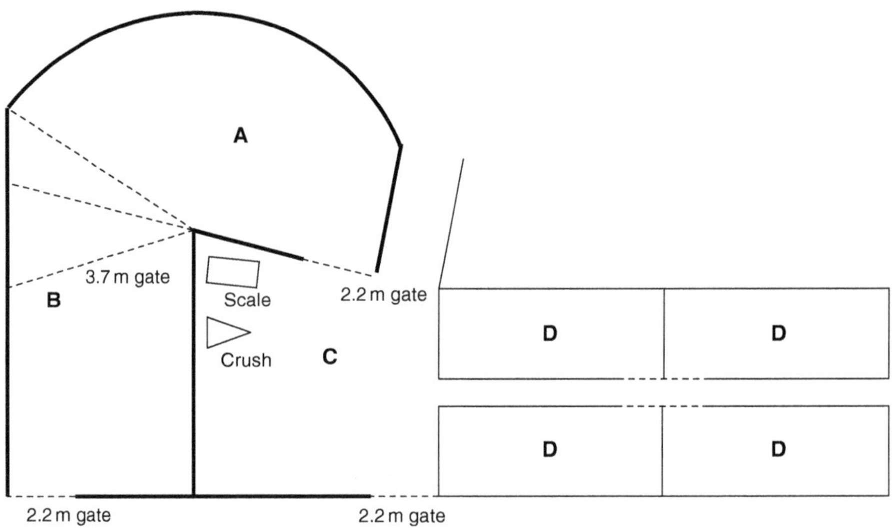

Fig. 10.7 Design for an ostrich handling facility, fitted with a revolving gate

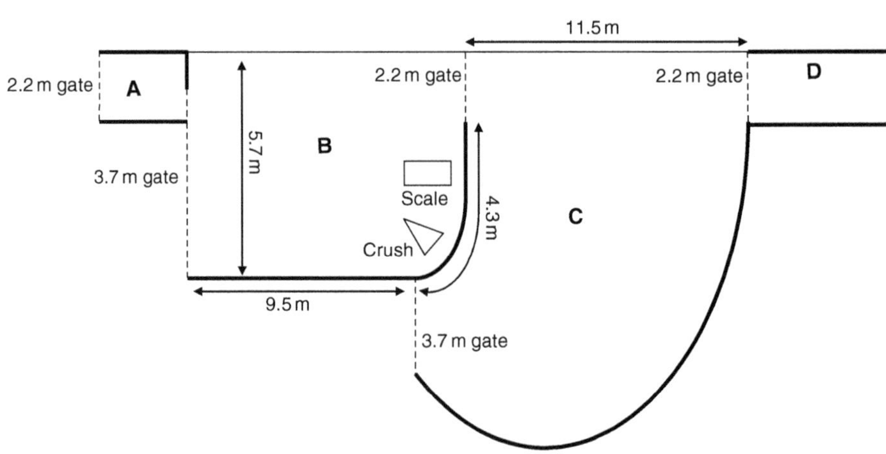

Fig. 10.8 Design of an ostrich handling facility, with a hospital kraal

Design 2

- This type of design is more suited for farms where birds need to be transported to e.g. camps that are far away in the veldt.
- Handling facility for ostriches that incorporates a dispatch area (A) from which ostriches are loaded onto the transport vehicle, a handling area (B) where all activities e.g. evaluations, scanning, and administration of medicines are performed. Adjacent to the handling area is the catching area (C), and a hospital camp (D) where sick or injured birds can be kept until treated. Broken lines indicate gates, and thick lines indicate where conveyor belting should be placed to minimise damage to birds during the handling process. The thin line represents normal camp fencing as is used for breeding camps (Fig. 10.8).

References

Alexander R McN, Maloiy GMO, Njau R, Jays AS (1979) Mechanics of running of the ostrich (*Struthio camelus*). J Zoo London 187:179–194

Anonymous (2004) The Meat Safety Act and Regulations. Act no. 40 of 2000, G.N.R. 8056. Johannesburg, South Africa: Lex Patria Publishers

Anonymous (2007) VPN/13/2007-01 Standards for *ante-mortem* and *post-mortem* meat inspection and hygiene control at ostrich meat establishments. [WWW document]. URL

Botha SSt.C, Hoffman LC, Britz TJ (2004a) Muscle pH and temperature changes in ostrich *M. iliofibularis* and *M. gastrocnemius, pars interna* during the first 24 hours *post-mortem*. In: Proceedings of the 2nd Joint Congress of the Grassland Society of Southern Africa and the South African Society of Animal Science. 28 June–1 July 2004. Goudini, South Africa, p 152

Botha SSt.C, Hoffman LC, Britz TJ, Nilsen BN, Slinde E (2004b) The effect of rigor-temperature on isometric tension, shortening and pH for ostrich *M. gastrocnemius, pars interna*. In: Proceedings of the 50th International Congress of Meat Science and Technology. August 2004. Helsinki, Finland, p 74

Burger WP, Peyrot P, Bekker A, Swart B, Theron LP, de Jesus AE, van Zyl E (1995) Microbial assessment of two methods of ostrich lairage. Report for Klein Karoo Co-op, Oudsthoorn, p 9

Chambers PG, Grandin T, Heinz G, Srisuvan T (2004) Effects of stress and injury on meat and by-product quality. In: Guidelines for Humane Handling, Transport and Slaughter of Livestock. FAO Corporate document repository. [WWW document]. URL. http://www.fao.org/DOCREP/003/X6909E/x6909e04.htm 28 August 2008

Crowther C, Davies R, Glass W (2003) The effect of night transportation on the heart rate and skin temperature of ostriches during real transportation. Meat Sci 64:365–370

Douglass A (1881) Ostrich farming in South Africa. Cassell, Petter, Galpin & Co, London, p 251

Engelbrecht A, Hoffman LC, Cloete SWP, van Schalkwyk SJ (2009) Ostrich leather quality: a review. Anim Prod Sci 49:549–557

Fasone V, Priolo A (2005) Effect of stress on ostrich meat quality. In: Proceedings of the 3rd International Ratite Science Symposium 14–16 October, Madrid, pp 393–396

Grandin T (1990) Design of loading and holding pens. Appl Anim Behav Sci 28:187–201

Grandin T (1991) Recommended animal handling guidelines for meat packers. American Meat Institute, Washington, DC

Hoffman LC (2005). A review of the Research conducted on Ostrich meat. In: Proceedings of the 3 rd International Ratite Science Symposium of the World's Poultry Science Association 14–16 Oct, pp 107–119

Hoffman LC, Botha S StC, Britz TJ (2006) Sensory properties of hot-deboned ostrich. Meat Sci 72:734–740

Hoffman LC, Cloete SWP, Van Schalkwyk SJ, SStC B (2009) Effect of bleeding method and low voltage electrical stimulation on meat quality of ostriches. S Afr J Anim Sci 39:209–213

Hoffman LC, Britz TJ, Schnetler DC (2010) Bruising on ostrich carcasses and the implications on the microbiology and losses in utilizable meat when removing them post-evisceration or post-chilling. Meat Sci 86(2):398–404

http://www.nda.agric.za/vetweb/, 28 August 2008

Huchzermeyer FW (1998) Diseases of ostriches and other ratites. Agricultural Research Council. Onderstepoort Veterinary Institute, South Africa, p 296

Kamau JM, Patrick BT, Mushi EZ (2002) The effect of mixing and translocating juvenile ostriches (*Struthio camelus*) in Botswana on the heterophil to lymphocyte ratio. Trop Anim Health Prod 34:249–256

Lambooij E, Potgieter CM, Britz CM, Nortjé GL, Pieterse C (1999a) Effects of electrical stunning methods on meat quality in ostriches. Meat Sci 52:331–337

Lambooij E, Pieterse C, Potgieter CM, Snyman JD, Nortjé GL (1999b) Some neural and behavioural aspects of electrical and mechanical stunning in ostriches. Meat Sci 52:339–345

Lambrechts H, Cloete SWP (1998) Activity budgets of adult breeding ostriches classified according to annual egg production of the previous breeding season. In: Animal production in Harmony with the Environment. In: Proceedings of the 36th National Congress of the South African Society of Animal Sciences, April 1998. University of Stellenbosch, pp 93–94

Liswaniso D, Purton MD, Boyd JS, Deeming DC (1996) Morphology of the distal region of the pelvic limb of the ostrich. In: Improving our understanding of Ratites in a Farming Environment. First International Scientific Ratite Congress, 27–29 March, 1996 Manchester, pp 9–10

McKeegan DEF, Deeming DC (1997) Effects of gender and group size on the time-activity budgets of adult breeding ostriches (*Struthio camelus*) in a farming environment. Appl Anim Behav Sci 51:159–177

Mitchell MA (1999) Welfare. In: Deeming DC (ed) The ostrich: biology, production, and health. CABI Publishing, Wallingford, pp 217–230

Morris CA, Harris SD, May SG, Jackson TC, Hale DS, Miller RK, Keeton JT, Acuff GR, Lucia LM, Savell JW (1995) Ostrich slaughter and fabrication. 1. Slaughter yields of carcasses and effects of electrical stimulation and post-mortem pH. Poult Sci 74:1683–1687

Paleari MA, Corsico P, Beretta G (1995) The ostrich: breeding, reproduction, slaughtering and nutritional value of the meat. Fleischwirtsch 75:1120–1123

Paleari MA, Camisasca S, Beretta G, Renon P, Corisco P, Bertolo G, Crivelli G (1998) Ostrich meat: physico-chemical characteristics and comparison with turkey and bovine meat. Meat Sci 3:205–210

Pfitzer S, Lambrechts H (2001) The use of haloperidol during the transport of adult ostriches. J S Afr Vet Assoc 72:2

Sabbioni A, Superchi P, Sussi C, Quarantelli A, Bracchi PG, Pizza A, Barbieri G, Beretti V, Zanon A, Zambini EM, Renzi M (2003) Factors affecting ostrich meat composition and quality. Ann Fac Med Vet Di Parm 23:243–252

Sales J, Mellett FD (1996) Post-mortem pH decline in different ostrich muscles. Meat Sci 42:235–238

SAOBC (2001) South African Ostrich Business Chamber (SAOBC) in conjunction with the National Council of Societies for the Prevention of Cruelty to Animals (SPCA) and the ARC – Animal Nutrition and Animal Products Institute. Code of Practice for the Transport, Handling and Slaughter of Ostriches

Schaefer AL, Jones SDM, Robertson WM, Brereton DA, Jeremiah LE (1995) Carcass yield and meat quality of ostriches under two different ante mortem management regimes. Final Report for the Canadian Ostrich Association, November 1995

Schaller NU, Herkner B, Prinzinger R (2005) Locomotor characteristics of the ostrich (*Struthio camelus*) I. Morphometric and morphological analyses. In: Proceedings of the 3 rd International Ratite Science Symposium 14–16 October, Madrid, pp 83–90

Severini M, Ranucci D, Miraglia D, Branciari R (2003) Preliminary study on the microbiological quality of ostrich (*Struthio camelius*) carcasses dressed in small Italian abattoirs. Ital J Food Sci 15:295–300

Van Schalkwyk SJ, Hoffman LC, Cloete SWP, Mellett FD (2005) The effect of feed withdrawal during lairage on meat quality characteristics in ostriches. Meat Sci 69:647–651

Wotton SB, Hewitt L (1999) Transportation of ostriches – a review. Vet Rec 145:725–731

Wotton S, Sparrey J (2002) Stunning and slaughter of ostriches. Meat Sci 60:389–394

Chapter 11
Ratite Conservation: Linking Captive-Release and Welfare

J.L. Navarro and M.B. Martella

Abstract Captive breeding for conservation purposes should promote a level of animal welfare that does not jeopardise post-release survival and reproduction in the wild. There are few documented cases of captive-release programmes in ratites; thus, pilot experiences in the Greater rhea *Rhea americana* and Lesser rhea *Rhea pennata* can be useful examples. In rheas, breeding should be carried out in large paddocks with grazing complemented by balanced commercial feed, keeping contact with humans and chicks to a minimum, and even leaving chicks under the care of an adult male. However, animals reared under the traditional intensive system (in smaller corrals and fed with processed food and chopped alfalfa) have shown acceptable acclimation, survival, and integration with their wild conspecifics when released into protected areas. Release should be carried out in the non-breeding season, using either a hard (performed upon arrival at the established area) or a soft (after an acclimation period in corrals located within the release area) approach. Because rheas are herbivorous and flightless birds, the selected site should be optimum in terms of vegetation type and structure. Clinical inspection and selection of young individuals without evident pathologies, careful loading and transportation in adequate crates, and a well-planned post-release monitoring are some other key factors. As captive-release is a promising tool for tackling conservation problems in ratites when other *in situ* actions are insufficient, the optimum captive conditions should emerge from the integration of the disciplines of animal welfare and conservation biology.

J.L. Navarro (✉)
Universidad Nacional de Córdoba – Consejo Nacional de Investigaciones Científicas y Técnicas, Centro de Zoología Aplicada, CC 122, Córdoba 5000, Argentina
e-mail: navarroj@efn.uncor.edu

M.B. Martella
Centro de Zoología Aplicada, Universidad Nacional de Córdoba, Rondeau 798, Córdoba 5000, Argentina
e-mail: martemo@efn.uncor.edu

P.C. Glatz et al. (eds.), *The Welfare of Farmed Ratites*, Animal Welfare 11,
DOI 10.1007/978-3-642-19297-5_11, © Springer-Verlag Berlin Heidelberg 2011

11.1 Introduction

In Argentina, populations of the two native ratites (the Greater Rhea, *Rhea americana*, and the Lesser Rhea, *Rhea pennata*) have decreased in the last years and both species have been categorised as "Near Threatened" by the International Union for Conservation of Nature and Natural Resources (IUCN 2010). The geographical distribution of the former species includes one of the most productive agricultural regions in the country; consequently, the Greater rhea is affected by the progressive reduction and fragmentation of the suitable habitats (Giordano et al. 2008). In the recent decades, natural grasslands and planted pastures for cattle raising have been replaced by annual crops (mainly soybean, *Glycine max*) because different local and global economic factors have turned agriculture into a more profitable activity than cattle production (Brown et al. 2005). Additionally, both rhea species are affected to a varying degree, depending on the region and the year, by poaching and illegal egg harvesting for local consumption (Bellis et al. 2004a, b, 2006; Barri et al. 2008). Wild populations of the Lesser Rhea, an inhabitant of the Patagonia steppes, are negatively affected by overexploitation and alteration of natural wetland meadows, locally known as *mallines* (flooded areas with the highest vegetation productivity in the region), for free range production of lambs (Bellis et al. 2006; Barri et al. 2008).

Assuming a probable need for reinforcing the diminished wild populations of rheas, researchers have conducted several experimental trials in the last decade, in which captive-bred individuals of both rhea species were released into selected sites within their natural geographical distribution (Bellis et al. 2004a, b; Navarro and Martella 2004; Martella and Navarro 2006; Juan et al. 2008).

11.1.1 The Captive-Release Approach to Ratite Conservation

The release of captive-bred animals into the wild for reinforcement or repopulation purposes usually has a wide spectrum of results, ranging from extremely successful ones to complete failures (Teixeira et al. 2007). Programmes currently using this strategy for conservation purposes of ratite birds are very scarce and include the release of 13 ostriches in Israel (Meyrav 2005) and ongoing actions for Kiwis (*Apteryx* spp.) (Colbourne et al. 2005). However, future releases are planned for the Ostrich (*Struthio camelus*) in Niger (Bishop et al. 2008; SCF 2010) and for a subspecies of the Lesser Rhea, the Andean or Puna Rhea (*Rhea pennata garleppi/tarapacencis*) in Perú (Navarro 2008). Also, there is a recent example of an accidental introduction of captive-reared ratites into a wild area far beyond their natural range: in northeast Germany three pairs of Greater Rheas escaped from a farm in 2000 and succeeded to establish and breed, founding a colony in Mecklenburg-Vorpommern. Currently, the size reached by this population is about a hundred birds (Deutsche Welle 2008; Korthals A and Philipp F in litt.).

Based on an analysis of the problems faced throughout the captive-release experiences of rheas in Argentina (see Navarro and Martella 2008), we have formulated recommendations that are linked to the animal's welfare, for increasing the chances of success when releasing captive-bred rheas into the wild. These recommendations could be also extrapolated to other ratites for which there are no specific data available.

11.1.2 Captive Breeding and Welfare

Understanding how to provide captive individuals with a sufficiently suitable environment to prevent welfare problems is usually attained by means of an analysis of animal welfare. The first difficulty encountered, however, is that welfare itself is not easily measured.

Because of the complex nature of welfare, it is generally necessary to use a number of variables to determine whether it has been achieved. Indeed, it is well known that under poor captive conditions animals are under severe stress (which is also difficult to measure, and causal factors vary among species and individuals) (Morgan and Tromborg 2007). Because of this negative association between stress and animal welfare, monitoring of glucocorticoid level has become a fairly common tool for measuring welfare. The use of glucocorticoids (corticosterone) as indicators of stress in ratites is a very recent approach; indeed, although stress has been indicated as an important cause of mortality in captive-bred species of this group, the only reports available are the studies of Mitchell et al. (1996) in ostriches and Lèche et al. (2009) in rheas.

The HPA system may be also activated by causes not necessarily related to stress; it is also possible that some of the factors compromising animal welfare do not activate HPA system. Therefore, the analysis of corticoid levels is usually strengthened with other measures, such as observing animal behaviour (a good indicator of how animals perceive environmental changes), particularly the frequency or intensity of depressive or aggressive and stereotyped behaviour. Another technique used to measure stress is monitoring signs of immunological alterations, syndromes, occurrence of injuries, disease, and morbidity (which are evident later than physiological or behavioural signs), or suppression of reproduction. In certain species, however, the occurrence of events or stimuli that generate short-lasting acute stress is beneficial, because this mechanism has reproduction-enhancing effects (Carlstead and Shepherdson 1994). Some authors have even proposed that, besides measuring biological functions and health of animals, it is necessary to include a way of measuring awareness, and feelings and emotions that animals can experience and, therefore, their suffering capacity (see review in Rushen 2003).

When animal breeding is aimed at reintroduction of individuals into the wild, some aspects of animal welfare can be pondered in different ways. It may be necessary to incorporate some stress-inducing stimuli or environmental conditions that would be suboptimal in the strict terms of the current welfare conditions, but that contribute to improve adjustment of animals to the real wild conditions they will be exposed to in the future (Håkansson and Jensen 2005; Mathews et al. 2005). Animal welfare is a pre-requisite to any *ex situ* conservation programme; if the programme includes translocation of individuals, however, during the captivity phase an appropriate welfare balance should be attained, so that survival and reproduction in the future phase of life in the wild is not threatened.

In the following sections we deal with animal welfare associated with captive-release projects for conservation purposes. As an example, we will refer to the case of rheas, showing several studies related to captive breeding and translocation of individuals and release into the wild; we discuss how the different management alternatives affect animal welfare conditions.

11.2 Producing the Individuals

Although rheas and their products have been traditionally used by men since before the Spanish colonisation in South America (Felce and Benarós 1943; Barbarán 2004), species of the ratite group are the ones that have been most recently incorporated into farming systems. Contradictorily, commercial breeding of rheas started in the USA and Europe, and only in the mid-1990s it was incorporated in its natural distribution range.

Artificial breeding of rheas currently poses the same problems usually encountered in any productive activity that has been recently introduced; however, these problems are accentuated by the fact that the species of this taxonomic group are quite different from the rest of birds, their distribution is restricted to South America, and they have been comparatively poorly studied. Therefore, there has been no research specifically addressing animal welfare in rheas, unlike in other farmed birds (including some ratites like the ostrich) (Mitchell 1999; Glatz 2005a, b). Although technicians recommend several alternatives and possible combinations for managing captive rheas (see Sarasqueta 1997; Vignolo 2006), their proposals lack rigorous scientific background. Thus, almost all reasonable alternatives are considered equally appropriate and, consequently, permitted by the corresponding inspection body. Finally, producers adopt what they consider the best alternative (Chang Reissig et al. 2001), without having objective information available on the effect of the most frequently used management and environmental enhancement measures. Therefore, research on welfare should be encouraged, using different methods, such as those proposed by Tarlow and Blumstein (2007), which they ranked from fitness indicators to disturbance indicators: breeding success, mate choice, fluctuating asymmetry, flight initiation distance, immunocompetence, glucocorticoids, and cardiac response.

Finally, standards regulating rhea captive breeding farms in Argentina (decrees 148/89 and 26/92, and provincial acts) are not directly oriented towards animal welfare, but mostly contemplate sanitary aspects and administrative issues related to authorisation and operation of captive breeding of wild animals in farms. Accordingly, regulations similar to the ones in force in New Zealand (AWAC 1998) should be applied in Argentina, which include animal welfare and provide more detailed information about good practises in captive breeding to attain welfare.

Besides the limitations mentioned above, there are a number of aspects that, although they may vary with animal age and among regions, should be taken into account specifically in captive breeding of rheas for further release in the wild.

11.3 Feeding

In semi-extensive breeding systems, where there are large spaces available, feeding is based on direct rotational grazing on natural or cultivated pastures (alfalfa, clover, chicory, or saltbush), complemented with commercial feed (any of the different types available in the market) (Fig. 11.1). In intensive systems, however, the standard diet consists mostly of commercial feed offered ad libitum [or based on consumption curves extrapolated from the ostrich; Navarro et al. (2000)], complemented with a variable proportion of packed or fresh (cut or chopped) plant material. Other components (soybean, probiotics) are occasionally incorporated in the diet, but their efficacy is not always supported by scientific research results (Huchzermeyer 1998; Dominino et al. 2006; Bazzano et al. 2007; Gri et al. 2008). Food is usually offered in one or more trays, bowls, or containers of different types or, most commonly in the case of plant matter, deposited directly on the ground. Rhea chicks prefer the feeders of the largest size available, independently of their colour (Vignolo 2006), whereas ostrich chicks exhibit a better feeding behaviour when feed is dispersed on the floor or in half-empty bowls than when food is presented in trays (Paxton et al. 1997).

Fig. 11.1 Greater rheas in a semi-extensive breeding system

There are no scientific studies indicating the appropriate composition of the diet and the specific balanced feed that meets rheas' nutritional requirements during the different developmental stages and seasons, or about the most suitable way to offer the food to the species. Preliminary studies with two balanced feeds available in the market indicated that adult rheas prefer some feeds for chickens to those that manufacturers claim to have formulated specifically for the species, and that there are also differences in preferences between sexes (Bazzano 2005). This result indicates the need to evaluate adaptation of rheas to environmental conditions based on objective tests rather than on inferences.

It is particularly important to avoid feeding the young rheas with an inadequate calcium:phosphorous ratio and inappropriately high levels of energy or protein (particularly when accompanied with low levels of exercise and frequent trauma to the proximal tibiotarsus). These conditions contribute to the development of different types of leg deformities encompassed under the denomination of bow-leg syndrome (Guittin 1986). This syndrome may also be caused by supplying drinking water that has an inadequate chemical composition. In Greater rhea, at least one case having this characteristic was detected in a commercial farm of the province of Córdoba (Argentina).

Some of these distortions of leg bones can develop rapidly, and can compromise locomotion and subsequent survival. This syndrome that is common in large free-living birds reared in captivity is one of the most important risk factors for death or euthanasia in ratites during the first 4 months following hatch (Moore 1996; Huchzermeyer 1998). Abnormal development of rhea chicks can be detected by monitoring their body mass regularly and comparing weight values with some baseline data for the species [e.g. Navarro et al. (2005)]. As the window of opportunity to implement corrective measures in the diet and husbandry practises is very narrow, early detection is paramount for obtaining successful results.

Whereas the intensive system is highly unnatural, captive (semi-extensive) or semi-captive systems that maintain the rheas mostly under direct grazing on pastures in large paddocks are more similar to the real situation in nature, and thus should be preferred for animals that will be released into the wild (Navarro and Martella 2002; Vignolo 2006) (Fig. 11.2). Additionally, Vignolo et al. (2001) reported that perinatal survival is higher in rheas bred using these methods than the classic intensive one (72% vs. 57%). However, rheas that were bred in comparatively small corrals and fed on processed feed and alfalfa (traditional intensive system) and later released in protected areas have shown fairly good acclimation, good survival rate, and integration with their wild conspecifics (Bellis et al. 2004a, b).

11.4 Chick Rearing and Filial Imprinting

In a captive-release programme, young should be produced under a system that minimises the contact and interaction with humans to avoid the birds becoming imprinted, very tame (naive) and, consequently, very prone to be hunted or captured in the wild.

Fig. 11.2 Greater rheas released into the wild

Chicks of these species are nidifugous, feed on their own, and remain under the care of an adult male from hatch to several months of age. During this period, the male provides heat, protection from predators, and guidance in sites where the food items are abundant. Because of the importance of the parent figure to newly born chicks, many intensive breeding farms have attempted to replace it with an older chick, some other animal (for example, rabbits or dwarf goats), or even with human presence in the paddocks for a long time, or with a human effigy or dummy (Huchzermeyer 1998; Navarro personal observations). Most of these approaches may reduce stress and lead to higher perinatal survival rates that can be attractive for commercial purposes. However, these practises also place disadvantages, especially regarding the reproductive behaviour of animals reaching the adult stage (for example, see Bubier et al. 1998; Huchzermeyer 1998), which are undesirable in captive-release programmes.

Rhea chicks (up to 3 or 4 months of age) that are reared under the traditional intensive method are usually maintained in brooder rooms with heating elements during night and cold or rainy weather, whereas they have free access to open areas (2–4 m^2 per bird) during the day (Fig. 11.3). Intensive rearing conditions are unnatural and some unfavourable environmental situations and wrong farming practises can lead to stress, which can affect behaviour, growth, perinatal survival, and reproduction (Vignolo et al. 2001; Navarro and Martella 2008). By contrast, there is some evidence in favour of the use of natural methods of incubation and chick rearing [including adoption, see Lábaque et al. (1999), Barri et al. (2005)], which would benefit chicks. By utilising these methods, one reduces the tameness of the animals and the occurrence of undesirable behaviours that may hamper the ability of the birds to adapt (survive and successfully breed) to their natural environment.

Fig. 11.3 Lesser rhea chicks in the intensive breeding system

Rearing rhea chicks separately by size or age (Sarasqueta 1997) may reduce transmission of endoparasites (Chang Reissig et al. 2001). The convenience of breeding chicks separately from juveniles of different sex or size has not been evaluated in terms of possible differences in aggression rate between systems. Meyer et al. (2003) studied this alternative in ostrich, and concluded that there was no need to raise animals of different sex separately. These conclusions, however, were solely based on economic aspects of production (number and location of kick mark scars on the skins, which yielded significant losses for producers). Had animal welfare also been considered, completely opposite recommendations would have been made.

11.5 Shelter

In the typical intensive breeding systems, chicks are enclosed in brooder houses during the night and on harsh weather days (Navarro and Martella 2002). Recommendations on dimensions, characteristics of the floor or bedding material, aeration, and ambient temperature have been mostly made on an empiric basis or by extrapolating data from other birds. Adult ratites, however, should not be kept indoors during the night or prolonged periods, because it can be counteractive in terms of animal welfare (T-AP 1997; Navarro personal observations). Rheas are adapted to fairly extreme climate conditions: Greater rhea inhabits regions where summer is characterised by strong insolation (animals usually seek shelter under the trees), and high temperature and humidity. Lesser rhea, in turn, is distributed in regions where winter is especially harsh, with extremely strong winds, very low temperatures, and important snowstorms, whereas summer is characterised by high insolation and extreme drought. The Andean subspecies is also adapted to high altitude conditions.

Along with the increasing significance that ratite farming has gained worldwide, there is growing interest in determining whether these animals' welfare is affected by adverse local climate conditions. There is particular concern about cold, wet weather, and the lack of sunshine, conditions that are far different to those prevailing in the natural environments (Deeming 1997, 1998; Horbanczuk 2002; Wöhr and Erhard 2005), because there is still some lack of knowledge about the appropriate husbandry conditions in each environment. In Argentina breeding and release into the wild of each rhea species are conducted only within the natural distribution range; in this way, such problems are avoided.

11.6 Incubation

Incubators that are especially designed for rheas, with automatic egg turning and forced air ventilation, should be used for efficient artificial incubation (Amelotti et al. 2007). However, natural incubation of manipulated clutches should also be considered an efficient method for producing chicks for release into the wild. Although some evidence indicates that artificial incubation does not alter the behaviour and survival of rhea chicks (Lábaque et al. 1999; Barri et al. 2005), whether it has beneficial effects on their welfare and subsequent fitness in the wild should be evaluated.

11.7 Genetic Management

In order to maintain the highest genetic similarity to conspecifics of the wild population to be reinforced, captive animals should not be objects of intended domestication or other ill-judged selection. However, genetic management of the breeding stock may be necessary for reducing inbreeding depression, loss of genetic diversity and genetic adaptation to captivity, and for eliminating genetic traits that would be lethal or would jeopardise reproduction in the wild (Frankham et al. 2004). This strategy requires some degree of deliberate selection by those breeding pairs supposed to be the fittest ones and their most advantageous combinations (by testing crosses). Additionally, these human interventions may have, in turn, an effect on animal welfare if social factors or animal's preferences are important for successful pair bonding in a given species (see Morgan and Tromborg 2007). As it has been proven that Greater rheas bred in commercial farms possess similar genetic variability to that of wild populations located in the same region, they can be an important source of genetic diversity for captive-release programmes (Alonso Roldán et al. 2010).

11.8 Ingestion of Harmful Objects

Stress triggered by desertion, disorientation, or frustration of birds that are deprived of, or unable to recognise, food may lead to compensatory behaviours such as disturbed eating, or feather or face pecking. These compensatory actions are easily disseminated, as they are started by one individual but can be rapidly copied by other conspecifics and soon adopted by the rest of the group. All ratite species in captivity are particularly prone to impaction of the proventriculus when they ingest abnormal amounts of non-digestible fibrous material (long stemmed alfalfa *Medicago sativa*, bermuda *Cynodon dactylon* or Kikuyu grass *Pennisetum clandestinum* stolons, or excessively long leaves), or non-food objects, such as metal nails, staples, mower blades, keys, syringe needles, wire, pieces of glass or metal sheet, wood shavings, large thorns or twigs, plastic or metallic objects, batteries, coins, mud, litter, sawdust, straw, soil, sand, stones that are large or with sharp edges, strings, cloth, etc. Impaction and hardware disease are associated with gastric stasis and perforation of the digestive tract by sharp objects, followed by septicemia or toxaemia that leads to mortality (Deeming and Dick 1995; Huchzermeyer 1998; Samson 1998; Chang Reissig and Robles 2001; Sen and Albay 2003; Aichinger 2007).

Foreign objects ingested by an animal may remain in the proventriculus and gizzard (even embedded in its wall) for a considerable period, without any evident symptoms, affecting the animal severely only when exposed to a stressful situation, such as transportation and release into a novel habitat (Deeming and Dick 1995). In general, the symptoms these birds may manifest are non-specific and to some extent may be confused with those of other conditions or diseases; hence, although diagnosis based on the use of standard metal detectors, a full clinical inspection including palpation of the proventriculus and auscultation of the gizzard, radiography or ultrasonography, and appropriate medical treatment can save affected birds, in most cases diagnosis is made during post-mortem examination. Therefore, rheas should be examined before being transported and, if possible, paddocks where rheas will be released should be checked to ensure all objects that could pose a threat if ingested by rheas because of their size or sharp or pointed characteristics are removed.

11.9 Handling

Similarly to the rest of ratites, rheas are fairly susceptible to undergo capture myopathy or exertional rhabdomyolysis. Rae (1992) suggests that it may be an acute manifestation of a subclinical problem of deficiency of vitamin E and possibly selenium. Inappropriate handling also usually produces respiratory and heart disease (Fowler 1995; Smith et al. 2005). Prolonged chases and restraints, particularly in ambient temperature above 27°C, are factors that usually trigger these problems, which may lead to acute and subacute mortality. Persons handling rheas should be trained to do so; indeed, these animals are dangerous due to the particular strength of their legs that makes them capable of delivering fast and powerful kicks and cause injury with their sharp toenails. As in the case of emus, rhea chicks and juveniles can be caught and handled by a single operator, but at least two persons are needed to restrain older animals in a relatively safe manner, so that a third person can perform a veterinary procedure (Fowler 1995; Raines 1998). The use of hoods facilitates handling of some ratites (Fowler 1995), but it is not usually helpful in handling rheas (personal observations).

11.10 Sexing and Identification

Sexing of animals is a very important aspect in appropriately forming the stock or the group of animals to be released. Sex can be positively determined using more or less laborious or invasive techniques that are traumatic to the animals, some of them being also more effective than others depending on the age of individuals. Adult Greater rheas can be sexed by their distinct dimorphic characters, particularly the plumage colour pattern (a character that is less evident in the Lesser rhea) and, to a lesser degree, by body size. This method, however, has several possibilities for error (even for experimented people), because some animals have intermediate characters, and are difficult to sex. Hence, accuracy of this type of sex determination method is considerably increased if conducted immediately before or during the reproductive season, when characters are more marked and especially because behavioural differences become fairly evident at that time.

Rheas of both species can be sexed with full accuracy at any age and season using a non-invasive molecular method, from DNA isolated from the calamus of feathers (Rossi Fraire and Martella 2006). This method should be preferred in terms of welfare, because it does not involve animal restraint.

Other somewhat more invasive (and less accurate) methods can be alternatively used to verify the presence or absence of a penis, such as internal digital palpation of the cloaca (Weeks and Bush 1974) or its eversion by external manipulation with the fingers, or its observation by the use of a nasal speculum (Brown and Kimbell 1972).

Rheas should be marked before being transported, preferably with marks suitable for the animal and that allow easy individual identification at a distance. A variety of visible external identifiers are used in ratites (Fig. 11.4), such as different types of collars (Martella and Navarro 1992), leg-bands or bracelets (placed either above or below the hock joint), neck or wing tags, or stock-marking paint, usually combined with a permanent electronic identification (a transponder, which is normally implanted in the muscle of the tail area of adults or in the piping muscle in the upper neck behind the head on the left of hatchlings) (Huchzermeyer 1998;

Fig. 11.4 Greater rhea marked with a numbered collar (*left*), and Lesser rhea with a leg-band (*right*)

CFIA 2006). In general, when appropriately placed, no evident problems are detected in marked individuals or in the rest of conspecifics sharing the pen with them (Lábaque 2006). However, no specific works have been documented confirming if any of these identification methods is more prone than others to generate welfare problems in rheas.

11.11 Health Management

Animal health status must be monitored on a daily basis, and veterinary consultation must be requested if a rapid change in an animal's condition, inappetence, lethargy, lameness, or any other signs of distress or disease are detected (refer Chap. 9).

Recommended treatments against microbes, which contribute most to ratite diseases and syndromes, are available in the literature and their efficacy is already known. However, the use of most of them has only little empirical basis; moreover, the poor empirical evidence available has resulted from studies on a single ratite species, or from extrapolation from those therapies successfully employed in other animals (even pertaining to other groups), because research in pharmacokinetics and pharmacodynamics has been conducted only in a few of the available drugs commonly administered to ratites (Abu-Basha et al. 2006, 2008; de Lucas et al. 2004, 2005a, b, 2008). This situation may lead to the popular and continuous use of some treatments or drugs that, inadvertedly, may have comparatively low efficiency for a given disease and to the use of incorrect doses or strength for an inadequate period. These practises, in turn, may contribute to the development of resistance to first-choice or broad-spectrum antimicrobials by infection agents. Some undesirable consequences of the inappropriate use of antimicrobials that can be faced in the next decades are prolonged illness and period of infectivity, diseases that will have no longer effective therapies, and greater risk of death.

When production is oriented to conservation, the animals must be sanitarily inspected regularly by a veterinarian, who must confirm that they are free of disease prior to release. Preventive administration of antimicrobials or medication in a captive-release programme is not encouraged because it not only contributes to the development of resistance by microorganisms, but also eventually interferes with the normal acquisition of immunity and other biochemical or physiological forms of resistance, tolerance or behavioural defences by the captive animals to parasites and pathogens that may be common in the wild (Wakelin 1996). Additionally, this practise will favour the survival of weaker animals, which cannot be easily identified, and will eventually mate and propagate their deleterious genes within the captive stock, resulting in an increasing number of unfit individuals. These animals, if released, will pose some risk of outbreeding depression to the existing wild population.

11.12 Beak Trimming

Beak trimming, rarely practised in Greater rheas, but frequently in Lesser rheas (Navarro personal observations), is necessary when deformity and overgrowth of the upper beak occur, which becomes progressively elongated and curved. As beak trimming is supposed to produce stress and chronic pain to the birds, it should be carried out by skilled operators that must follow strict guidelines of veterinary surgery protocols.

This specific deformity of the beak is of unknown aetiology and may be associated with insufficient wear and tear or nutritional deficiencies, but may also have a hereditary and/or congenital basis. Therefore, birds exhibiting this problem should not be allowed to reproduce within any stock that is raised with the primary objective of producing individuals destined for release into wild areas.

11.13 Transport

Few scientific studies have been undertaken to compare the stress or other effects on welfare caused by different handling and transport procedures or techniques (Mitchell et al. 1996; Wotton and Hewitt 1999; Minka and Ayo 2007). Some recommendations have been provided, but mostly premised on empirical basis (Huchzermeyer 1998; Wotton and Hewitt 1999). These recommendations have served as the basis for formulating a number of standards in several countries on ratite loading and transport (EFSA 2004). There are no studies indicating how rheas of different ages are affected by type and time of transport usually used. There is only evidence of the cases that were deemed successful and of those that failed completely or partly because of manifest handling errors during transport. In the latter cases, the most frequent problems were due to excessive chasing for capturing rheas before transport or during loading, or because of the slippery floor of the container used, and/or careless driving or too long stops during the journey, or because of the use of an inappropriate physical restraint (e.g. animals transported with their legs tied together, despite the comparatively short transport journey, approximately 12 km). In these cases the affected rheas suffered important injuries (fractures, lacerations, or capture myopathy) and high mortality. They usually died suddenly – within a few hours – or several days after transport (Navarro JL unpublished data).

The personnel involved in the capture and transport of rheas should be instructed appropriately about the correct way to manipulate, load, transport, and release these animals. Rheas must be herded slowly and quietly and not be chased in excess because they become too agitated, stressed and uneasy, either before or during transportation. Loading into trailers should be made either through horizontal bays or very shallow incline stable ramps with non-slippery flooring, or by catching and gently hand-lifting one individual at a time. Carriage over short or long distances (up to 12 h in transit) should be conducted either in specially built trailers, or in those for one or two horses or for livestock. Rheas can be also transported inside crates that can be loaded and then placed in the back of a pickup truck. The birds can be loaded individually or, preferably, in pairs – selecting animals of the same size – in crates of adequate dimensions (they must have sufficient space to sit down), with flooring or bedding that prevents slipping and trauma, and enough vents to ensure good air flow, but in which vision to the exterior is blocked. In our case, wooden crates of 1.5 m high × 1.5 m long × 0.90 m wide, and a layer of 3–4 cm of hay or sand, or a woven mat, with many holes (2-cm diameter) drilled at the upper part of lateral panels, were successfully used several times for journeys of approximately 400 km. Upon arrival at the release site, the crates were unloaded individually by two to three workmen, and once each door was open, the rheas walked off the crate unaided.

Rheas should be loaded and transported in hours of non-extreme temperatures, and if possible under dim light, early in the morning or in the evening. During the journey to the release area, driving should be careful, and provisions have to be made to minimise the frequency and duration of stops. Provincial and national regulations regarding transportation of this type of animals must be strictly followed.

11.14 Planning the Release

Individuals selected for release should be normal in all aspects (do not show leg, toe, or beak deformities) and exhibit good condition, not be remarkably tame, and be between a minimum of 10 and a maximum of about 18 months of age. Rheas within this age range are easier to manipulate and transport and, consequently, less prone to suffer injuries that in some cases may affect their succeeding survival and reproduction. This aspect should be very thoroughly considered in the case of males, which are usually more vigorous and aggressive towards humans and male conspecifics than females. Although release has never been performed in rheas of younger ages, the comparatively high mortality that usually occurs during the first six months of life in chicks under captive conditions (Navarro and Martella 2002) suggests that individuals under 10 months of age will have fewer chances to succeed in the wild.

Within the areas of high suitability, those exhibiting the highest level of protection should be selected. Naturally protected areas, where access is controlled and therefore poaching is nearly absent, should be prioritised for the release.

The most appropriate site for release should be selected in advance, based on the highest suitability of the available habitats for satisfying the life requirements of the species (i.e. food, reproduction, and cover). Habitat/s should be always assessed by an objective method (Stamps and Swaisgood 2007). A practical method for assessing species–habitat relationship for Greater rheas based on satellite image texture has been recently documented (Bellis et al. 2008). Habitat suitability should be assessed even if there is a natural population of the species in the site, because this population could be currently facing adverse circumstances. In such a case, those circumstances must be identified, quantified, and managed to reduce, neutralise, or eliminate their negative action, before starting reinforcement with new animals. If the particular anthropogenic stressors that negatively affect the wild population cannot be managed *in situ* in any way, and there is no chance for selecting other site for release, captive animals should be properly trained to cope with those adverse conditions.

Deficiencies in critical behaviours that could reduce the ability of rheas to survive and reproduce in nature should be examined. It is especially important that appropriate anti-predatory, social, and exploratory behaviours be developed by the candidate individuals for release in such a new and variable environment (Håkansson and Jensen 2005). When possible, naive rheas should be trained for predator recognition and avoidance before being released into the wild (Schetini de Azevedo and Young 2006a, b). In areas where occurrence of predators (including dogs) is suspected, some exclusion measures for these animals should be implemented at the release site.

The release should ideally take place well before the start of the breeding season to allow rheas to become acclimated to the novel environment before reproduction. In males this approach will decrease the probability of fights with wild males, whereas in females it will impede the eventual breaking of eggs in formation as a consequence of accidental hits during transport.

From a cost/effectiveness point of view, a "hard" release method where animals are released just upon arrival at the established area is recommended. In both rhea species, there were no differences in success between this method and the "soft" one, where individuals supposedly are acclimated progressively to their novel environment by confining them for long periods in corrals located within the area selected for release (Bellis et al. 2004a, b).

The movements and activities of the animals should be monitored for some time post-release. For this purpose, individual marks that can be distinguished from a distance with the aid of binoculars or telescopes should be tested well in advance

Fig. 11.5 Greater rhea with a radio-transmitter mounted in an auto detachable self-expansible collar

(Martella and Navarro 1992). Ideally, some of these birds will be provided with a radiotelemetry device that facilitates locating them in the field. Radio-transmitters with batteries that provide two years of service life, mounted in auto detachable self-expansible bracelets or collars are a practical option for these large birds (Fig. 11.5). This type of mounting readjusts itself if the animal continues growing after release, and increases chances of survival if the bird gets hooked by the collar to a thorny bush or a barbed wire. Although this characteristic also increases chances of losing the collar, it could be located with the receiver (provided that the battery and the transmitter work), and then refurbished and reutilised in another bird.

The protocol for reintroduction of wild species IUCN (1998) must be cautiously followed in all cases.

11.15 Release Procedure and Post-Release Monitoring

General recommendations and regulations for the transport of this ratite should be followed.

Monitoring of the animals should start immediately after their release into the wild. The first monitoring period should be continuous (several times a day) and as long as possible (not less than two weeks). When monitoring is interrupted, even for a couple of days, there is a chance of losing the radio signal (at least from land) and this could lead to a very tedious search (sometimes unsuccessful) of the lost animals. This situation is particularly valid when the size of the home range of the species is unknown, or if one presupposes that it may be large.

In the previous case, or if the monitoring period suffers many or long interruptions, radio tracking from aircrafts or satellites should be considered as the main strategy or a complementary one.

The release of captive-produced individuals into natural habitats seems to be a promising tool for recovering threatened ratite populations when other more conservative *in situ* actions are insufficient to ensure the preservation of the target species. However, as it is a controversial strategy, it should be carried out with extreme care, combined with other conservation measures (e.g. law enforcement, habitat protection, and public awareness raising), and should include monitoring of the birds released. The documentation and dissemination of meaningful information and lessons learned throughout the process will be imperative for drawing sound conclusions about the most effective rearing and release protocols for each species.

11.16 Concluding Remarks

Further scientific research linking conservation and welfare is necessary to contribute to improve the current knowledge and, basically, to test critical hypotheses and provide more accurate predictions about the relations between the different components of the environment and welfare of rheas. Of particular importance is the evaluation of phenotypic plasticity of the species in response to environmental change, especially considering how rapidly it is taking place because of both land use and global climate change.

Promoting animal welfare through recommendations on management of captive environments based on the systematic analysis of evidence requires availability of an adequate source of scientific articles and reports. According to Pullin and Knight (2009), this database could be implemented in a relatively simple manner thanks to the current communication technologies, if the scientific community is adequately encouraged. In this way we will be able to determine the actual effectiveness of the measures usually disseminated among producers, or those measures generated under pressure exerted by NGOs, and/or those imposed by organisations in charge of inspecting farms and breeding centres.

Public dissemination of the deficiencies in the basic knowledge of welfare indicators in this (and other) species is, despite the entailed risks, an honest and necessary attitude to increase reliability on, and support to, scientists by groups concerned with animal care and protection and by the rest of society. If scientists fail to express this self-criticism, responding to the growing pressure imposed by social awareness about animal welfare in captive management systems will be increasingly difficult. Thus, it may be demonstrated that measures recommended to improve the conditions where animals are reared, if based on scientific grounds rather than on speculations, will have positive and significant effects on welfare.

References

Abu-Basha EA, Idkaidek NM, Hantash TM (2006) Pharmacokinetics and bioavailability of doxycycline in ostriches (*Struthio camelus*) at two different dose rates. J Vet Sci Suwon-si, Korea 7(4):327–332

Abu-Basha E, Gehring R, Hantash TM, Al-Shunnaq AF, Idkaidek NM (2008) Pharmacokinetics and bioavailability of sulfadiazine and trimethoprim following intravenous, intramuscular and oral administration in ostriches (*Struthio camelus*). J Vet Pharm Ther 32:258–263. doi:10.1111/j.1365-2885.2008.01036.x

Aichinger A (2007) Impactação de proventrículo e ventrículo por corpos estranhos em avestruzes. Rev Acad, Curitiba 5(3):311–315

Alonso Roldán V, Navarro JL, Gardenal CN, Martella MB (2010) May captive populations of Greater Rhea (*Rhea americana*) act as genetic reservoirs in Argentina? Zoo Biol. doi:10.1002/zoo.20314

Amelotti I, Martella MB, Navarro JL (2007) Difiere la probabilidad de eclosión de los huevos de ñandú (*Rhea americana*), según su ubicación dentro de la incubadora? Rev Argent Prod Anim 27:41–46

AWAC (1998) Welfare Advisory Committee, Ministry of Agriculture and Forestry, New Zealand Government. Code of Animal Welfare 21

Barbarán FR (2004) Usos mágicos, medicinales y rituales de la fauna en la Puna del Noroeste Argentino y Sur de Bolivia. Contrib Manejo Vida Silv Latinoam 1:1–26

Barri FR, Navarro JL, Maceira NO, Martella MB (2005) Rearing Greater Rhea (*Rhea americana*) chicks: is adoption more effective than the artificial intensive system? Br Poult Sci 46:22–25. doi:10.1080/00071660400023888

Barri FR, Martella MB, Navarro JL (2008) Effects of hunting, egg harvest and livestock grazing intensities on density and reproductive success of lesser rhea *Rhea pennata pennata* in Patagonia: implications for conservation. Oryx 42:607–610. doi:10.1017/S0030605307000798

Bazzano G (2005) Evaluación del crecimiento y del éxito reproductivo del ñandú (*Rhea americana*) bajo el efecto de diferentes dietas. MSc Thesis, Universidad Nacional de Córdoba, Córdoba, Argentina

Bazzano G, Navarro JL, Martella MB (2007) Effects of different diets on growth and survival of Greater Rhea (*Rhea americana*) chicks. Archiv fur Geflugelkunde 71:117–121

Bellis LM, Martella MB, Navarro JL (2004a) Habitat use by wild and captive-reared Greater Rheas in agricultural landscapes. Oryx 38:304–310. doi:10.1017/S0030605304000547

Bellis LM, Martella MB, Navarro JL, Vignolo PE (2004b) Home range of Greater and Lesser Rhea in Argentina: relevance to conservation. Biodivers Conserv 13:2589–2598. doi:10.1007/s10531004-1086-0

Bellis LM, Navarro JL, Vignolo P, Martella MB (2006) Habitat preferences of lesser rhea in Argentine Patagonia. Biodivers Conserv 69:90–93

Bellis LM, Pidgeon AM, Radeloff VC, St-Louis V, Navarro JL, Martella MB (2008) Modelling habitat suitability for Greater Rheas based on satellite image textures. Ecol Appl 18(8):1956–1966

Bishop K, Black P, Fleischer R, Hallager S, Houston B, Mace M, Monfort S, Newby J, Tidmus S (2008) Saving the World's: the Ostrich recovery project in Niger. CONNECT Magazine, September http://www.aza.org/Publications/2008/09/f_ostrich_sep08_web.pdf2008. Accessed 30 Dec 2008

Brown A, Martínez Ortiz U, Acerbi M, Corchera J (2005) Situación ambiental argentina, 3rd edn. Fundación Vida Silvestre Argentina, Buenos Aires

Brown DA, Kimbell RH (1972) Maintaining rheas *Rhea americana* at Forth Worth Zoo. Int Zoo Yearb 12:187–189

Bubier NE, Paxton CGM, Bowers P (1998) Deeming DC: Courtship behaviour of ostriches (*Struthio camelus*) towards humans under farming conditions in Britain. Br Poult Sci 39:477–481

Canadian Food Inspection Agency (CFIA)(2006). Identification. In: Accredited Veterinarian's Manual. http://www.inspection.gc.ca/english/anima/heasan/man/avmmva/avmmva_mod2e.shtml#a2.1. Accessed 8 Mar 2011
Carlstead K, Shepherdson DJ (1994) Effects of environmental enrichment on reproduction. Zoo Biol 13(5):447–459
Chang Reissig E, Robles CA (2001) Gizzard impaction in Lesser Rhea chicks (*Pterocnemia pennata*) raised on farms in Patagonia, Argentina. Avian Dis 45:240–244
Chang Reissig E, Martella MB, Navarro JL, Robles CA (2001) Aspectos sanitarios de la cría del Choique (*Pterocnemia pennata*) en granjas de la Patagonia Argentina. Rev Med Vet 82(6):324–326
Colbourne R, Bassett S, Billing T, McCormick H, McLennan J, Nelson A, Robertson H (2005) The development of operation nest egg as a tool in the conservation management of kiwi. Sci Conserv 259:24
de Lucas JJ, Rodríguez C, Waxman S, Gonzalez F, de Vicente ML, San Andres MI (2004) Pharmacokinetics of enrofloxacin after single intravenous and intramuscular administration in young domestic ostrich (*Struthio camelus*). Search Results. J Vet Pharm Ther 27(2):119–122
de Lucas JJ, Rodríguez C, Martella MB, Lábaque MC, Navarro JL, San Andrés MI (2005a) Pharmacokinetics of enrofloxacin following intravenous administration to greater rheas: a preliminary study. Res Vet Sci 78:265–267
de Lucas JJ, Rodríguez C, Waxman S, Gonzalez F, Uriarte I, San Andrés MI (2005b) Pharmacokinetics of marbofloxacin after single intravenous and intramuscular administrations in young domestic ostrich. Vet J 170:364–368
de Lucas JJ, Navarro JL, Rubio S, Vignolo PE, Asís VC, González F, Rodríguez C (2008) Pharmacokinetic behaviour of enrofloxacin in greater rheas following a single-dose intramuscular administration. Vet J 175:136–138
Deeming DC (1997) Effect of climatic conditions on the behaviour of adult ostriches (*Struthio camelus*) in Britain. Anim Welf 6(4):349–356
Deeming DC (1998) Effect of winter climatic conditions on the behaviour of adult ostriches (*Struthio camelus*) on a British farm. Anim Welf 7(3):307–315
Deeming DC, Dick ACK (1995) Ingestion of metal objects by ostriches (*Struthio camelus*). Vet Rec 137:99–100
Deutsche Welle (2008) Los ñandúes se acomodan en el norte de Alemania. http://www.dw-world.de/dw/article/0,7580,3505065,00.html Ecología. 22 July. Accessed 19 August 2010
Dominino J, Maestri DM, Lábaque MC, Martella MB, Navarro JL (2006) Hay mejor eclosión, supervivencia y crecimiento perinatal, adicionando soja a la dieta de los ñandúes reproductores? Rev Argent Prod Anim 26:1–9
European Food Safety Authority (EFSA)(2004) The welfare of animals during transport. Scientific Report of the Scientific Panel on Animal Health and Welfare on a request from the Commission related to the welfare of animals during transport (Question N° EFSA-Q-2003-094). European Food Safety Authority. http://www.efsa.europa.eu/cs/BlobServer/Scientific_Opinion/ahaw_report_animaltransportwelfare_en1.pdf?ssbinary=true2004. Accessed 19 Aug 2010
Felce E, Benarós L (1943) El Ñandú o Avestruz americano: su vida – sus costumbres. Colección Flor, Dovile, Buenos Aires
Fowler ME (1995) Restraint and handling of wild and domestic animals, 2nd edn. Wiley-Blackwell, New York
Frankham R, Ballou JD, Briscoe DA, McInnes KH (2004) A primer of conservation genetics. Cambridge University Press, Cambridge
Giordano PF, Bellis LM, Navarro JL, Martella MB (2008) Abundance and spatial distribution of Greater Rhea *Rhea americana* in two sites of the pampas grasslands with different land use. Bird Conserv Int 18:63–70. doi:10.1017/S0959270908000075
Glatz P (2005a) Effect of declawing on behaviour and skin quality in ostriches. In: Proceedings of the 3rd Ratite Science Symposium of the World's Poultry Science Association and XII World Ostrich Congress, Madrid, Spain, pp 157–162

Glatz P (2005b) Best practice methods to declaw ostriches to improve skin quality. In: Proceedings of the 3rd Ratite Science Symposium of the World's Poultry Science Association and XII World Ostrich Congress, Madrid, Spain, pp 397–400

Gri L, Martella MB, Navarro JL (2008) Efficacy of probiotics in intensively reared Greater Rhea (*Rhea americana*) chicks. Archiv fur Geflugelkunde 72:132–139

Guittin P (1986) Bow leg syndrome in Ratite birds. Avic Mag 92(2):70–79

Håkansson J, Jensen P (2005) Behavioural and morphological variation between captive populations of red junglefowl (*Gallus gallus*) – possible implications for conservation. Biol Conserv 122:431–439

Horbanczuk JO (2002) The ostrich. Warsaw, Poland

Huchzermeyer FW (1998) Diseases of ostriches and other ratites. Agricultural Reserch Council, Onderstepoort Veterinary Institute, Onderstepoort

IUCN (1998) Guidelines for re-introductions. IUCN/Species Survival Commission Re-introduction Specialist Group, Cambridge

IUCN (2010) IUCN red list of threatened species. Version 2010.2. http://www.iucnredlist.org. Accessed 19 Aug 2010

Juan EE, Bazzano G, Navarro JL, Martella MB (2008) Area de acción y desplazamientos de Ñandúes silvestres y provenientes de cautiverio en pastizales del centro de Argentina. In: Resúmenes de la XII Reunión Argentina de Ornitología, San Martín de los Andes, Argentina, pp 132

Lábaque MC (2006) Productividad, adaptación y conservación del Ñandú (*Rhea americana*). PhD Thesis, Universidad Nacional de Córdoba, Córdoba, Argentina

Lábaque MC, Navarro JL, Martella MB (1999) A note on chick adoption: a complementary strategy for rearing rheas. Appl Anim Behav Sci 63:165–170. doi:10.1016/S0168-1591(99) 00006-4

Lèche A, Busso JM, Hansen C, Navarro JL, Marín RH, Martella MB (2009) Physiological stress in captive Greater rheas (*Rhea americana*): highly sensitive plasma corticosterone response to an ACTH challenge. Gen Comp Endocrinol 162:188–191. doi:10.1016/j.ygcen.2009.03.014

Martella MB, Navarro JL (1992) Capturing and marking Greater Rheas. J Field Ornithol 63:117–120

Martella MB, Navarro JL (2006) Proyecto Ñandú. Manejo de *Rhea americana* y *R. pennata* en Argentina. In: Bolkovic ML, Ramadori DE (eds) Manejo de Fauna en Argentina: Proyectos de Uso Sustentable, Dirección de Fauna Silvestre. Secretaría de Ambiente y Desarrollo Sustentable, Buenos Aires, pp 39–50

Mathews F, Orros M, McLaren G, Gelling M, Foster R (2005) Keeping fit on the ark: assessing the suitability of captive-bred animals for release. Biol Conserv 121:569–577

Meyer A, Cloete SWP, Brown CR (2003) The influence of separate-sex rearing on ostrich behaviour and skin damage. S Afr J Anim Sci 33(2):95–104

Meyrav J (2005) History in the making in southern Israel – reintroduction of ostriches into the wild. http://www.birdingisrael.com/birdNews/special_reports/ostriches.htm. Accessed 19 Aug 2010

Minka NS, Ayo J (2007) Road transportation effect on rectal temperature, respiration and heart rates of ostrich (*Struthio camelus*) chicks. Vet Arhiv 77:39–46

Mitchell MA (1999) Welfare. In: Deeming DC (ed) The ostrich: biology, production and health. CABI, Wallingford, pp 217–230

Mitchell MA, Kettlewell PJ, Sandercock DA, Maxwell MH, Spackman D (1996) Physiological stress in ostriches during transportation. In: Deeming DC (ed). Improving our understanding of Ratites in a farming environment, Ratite Conference, Oxfordshire, pp 79–80

Moore SJ (1996) The performance of farmed ostrich chicks in eastern Australia. Prev Vet Med 29:91–106

Morgan KN, Tromborg CT (2007) Sources of stress in captivity. Appl Anim Behav Sci 102:262–302

Navarro JL (2008) Curso-Taller Manejo de Ñandúes (Suris). In: Lescano Rivero AL, Díaz Ceballos C, Flores Cuba L, Vilca Ticona JL (eds) Proyecto Especial Binacional Lago Titicaca,

Dirección de Desarrollo Agrícola y Medio Ambiente, Instituto Nacional de Desarrollo, Ministerio de Agricultura del Perú, Puno (Perú), p 31
Navarro JL, Martella MB (2002) Reproductivity and raising of Greater Rhea (*Rhea americana*) and Lesser Rhea (*Pterocnemia pennata*) – a review. Archiv fur Geflugelkunde 66:124–132
Navarro JL, Martella MB (2004) Conservación y manejo de ñandúes en Argentina. V Jornada Nacional de Ornitología del Perú Dilloniana 3:5–7
Navarro JL, Martella MB (2008) The relevance of captive breeding to conservation of native ratites in Argentina: an overview. Aust J Exp Agric 48:1302–1307
Navarro JL, Bellis L, Lábaque MC, Martella MB (2000) Crecimiento en pichones de Choique en criaderos: implicancias en el consumo y costos de alimentación. In: Robles C, Navarro JL (eds) Actas del Taller sobre Conservación y manejo del Choique en Patagonia. Bariloche, Argentina, pp 5–7
Navarro JL, Vignolo PE, Demaría MR, Maceira NO, Martella MB (2005) Growth curves of farmed Greater Rheas (*Rhea americana albescens*) from central Argentina. Archiv fur Geflugelkunde 69(2):90–93
Paxton CGM, Bubier NE, Deeming DC (1997) Feeding and pecking behaviour in ostrich (*Struthio camelus*) chicks in captivity. Br Poult Sci 38:151–155
Pullin AS, Knight TM (2009) Doing more good than harm: building an evidence-base for conservation and environmental management. Biol Conserv 142:931–934
Rae M (1992) Clinical anatomy of Ratites. Proceedings from the Annual Conference of the Association of Avian Veterinarians, pp 328–325
Raines AM (1998) Restraint and housing of ratites. In: Shane SM, Tully TN (eds) The veterinary Clinics of North America Food Animal Practice. Ratites. W Saunders Co, Philadelphia, pp 387–400
Rossi Fraire HJ, Martella MB (2006) DNA test applied to sex the Lesser Rhea (*Rhea pennata pennata*). Br Poult Sci 47(3):375–377
Rushen J (2003) Changing concepts of farm animal welfare: bridging the gap between applied and basic research. Appl Anim Behav Sci 81:199–214
Sahara Conservation Fund (SCF) (2010) Ostrich Conservation, Niger. http://www.saharaconservation.org/spip.php?page=project&id_article=36. Accessed 19 Aug 2010
Samson J (1998) Behavioral problems of farmed Ostriches in Canada. Can Vet J 37:412–414
Sarasqueta DV (1997) Cría de ñandúes. In: Carbajo García E, Castello Fontova F, Castello Llobet JA, Gurri Lloveras A, Marín M, Mesiá García J, Sales J, Sarasqueta DV (eds) Cría de Avestruces, Emúes y Ñandúes. Real Escuela de Avicultura, Barcelona, pp 327–363
Schetini de Azevedo C, Young RJ (2006a) Behavioural responses of captive-born Greater Rheas *Rhea americana* Linnaeus (Rheiformes, Rheidae) submitted to antipredator training. Rev Bras Zool 23:186–193
Schetini de Azevedo C, Young RJ (2006b) Do captive-born Greater Rheas *Rhea americana* Linnaeus (Rheiformes, Rheidae) remember antipredator training? Rev Bras Zool 23:194–201
Sen S, Albay MK (2003) Fatal impaction of the stomach in two farmed ostriches. Irish Vet J 56(6):317–318
Smith KM, Murray S, Sanchez C (2005) Successful treatment of suspected exertional myopathy in a rhea (*Rhea americana*). J Zoo Wild Med 36(2):316–320
Stamps JA, Swaisgood RR (2007) Someplace like home: experience, habitat selection and conservation biology. Appl Anim Behav Sci 102:392–409
T-AP (1997) Standing committee of the European convention for the protection of animals kept for farming purposes (T-AP): Recommendation concerning Ratites (Ostriches, Emus and Rheas) adopted by the Standing Committee on 22 Apr 1997. http://www.coe.int/t/e/legal_affairs/legal_co-operation/biological_safety%2C_use_of_animals/farming/rec%20ratites%20e.asp#P182_11578. Accessed 19 Aug 2010
Tarlow EM, Blumstein DT (2007) Evaluating methods to quantify anthropogenic stressors on wild animals. Appl Anim Behav Sci 102:429–451

Teixeira CP, de Azevedo CS, Mendl M, Cipreste CF, Young RJ (2007) Revisiting translocation and reintroduction programmes: the importance of considering stress. Anim Behav 73:1–13

Vignolo PE (2006) Análisis comparativo costo – efectividad de dos sistemas de cría del ñandú (*Rhea americana*) diseñados para la introducción de ejemplares en la vida silvestre. MSc Thesis, Universidad Nacional de Córdoba – Universidad Nacional de Río Cuarto, Córdoba, Argentina

Vignolo PE, Martella MB, Navarro JL, Maceira NO, Demaria MR (2001) Preferencia alimentaria y tasa de crecimiento en pichones de ñandú moro (*Rhea americana*). Rev Argent Prod Anim 21:9–16

Wakelin D (1996) Immunity to parasites: how parasitic infections are controlled, 2nd edn. Cambridge University Press, Cambridge

Weeks SE, Bush M (1974) Sexing ratites. Int Zoo Yearb 14:141–142

Wöhr A, Erhard M (2005) Ostrich farming in Germany – an animal welfare issue? In: Proceedings of the 3rd Ratite Science Symposium of the World's Poultry Science Association and XII World Ostrich Congress, Madrid, Spain, pp 145–156

Wotton SB, Hewitt L (1999) Transportation of ostriches – a review. Vet Rec 145(25):725–731

Index

P.C. Glatz et al. (eds.), *The Welfare of Farmed Ratites*, Animal Welfare 11, DOI 10.1007/978-3-642-19297-5, © Springer-Verlag Berlin Heidelberg 2011

www.ingramcontent.com/pod-product-compliance
Ingram Content Group UK Ltd.
Pitfield, Milton Keynes, MK11 3LW, UK
UKHW022212051026
15624UKWH00001B/130
9783642192968